# bioterror and biowarfare

## a beginner's guide

From anarchism to artificial intelligence and genetics to global terrorism, Beginner's Guides equip readers with the tools to fully understand the most challenging and important debates of our age. Written by experts in a clear and accessible style, books in this series are substantial enough to be thorough but compact enough to be read by anyone wanting to know more about the world they live in.

Other titles available in this series:

**anarchism**
ruth kinna

**anti-capitalism**
simon tormey

**artificial intelligence**
blay whitby

**the brain**
ammar al-chalabi, martin r. turner & r. shane delamont

**democracy**
david beetham

**energy**
vaclav smil

**evolution**
burton s. guttman

**evolutionary psychology**
robin dunbar, louise barrett & john lycett

**genetics**
anthony griffiths, burton guttman, david suzuki & tara cullis

**global terrorism**
leonard weinberg

**NATO**
jennifer medcalf

**the palestine–israeli conflict**
dan cohn-sherbok & dawoud el-alami

**postmodernism**
kevin hart

**quantum physics**
alastair i. m. rae

**religion**
martin forward

Forthcoming:

**the small arms trade**
matthew schroeder, rachel stohl & daniel m. smith

**biodiversity**
john spicer

**capitalism**
andrew kilmister & gary browning

**criminal psychology**
ray bull

# bioterror and biowarfare

## a beginner's guide

malcolm dando

ONEWORLD
OXFORD

bioterror and biowarfare: a beginner's guide

Oneworld Publications
(Sales and editorial)
185 Banbury Road
Oxford OX2 7AR
England
www.oneworld-publications.com


A CIP record for this title is available
from the British Library

ISBN-13: 978–1–85168–447–2
ISBN-10: 1–85168–447–6

Typeset by Jayvee, Trivandrum, India
Cover design by the Bridgewater Book Company
Printed and bound by WS Bookwell, Finland

# contents

# list of tables

# list of figures

# preface

In late 2001, just after the attacks by hijacked aircraft in the United States, several anthrax-impregnated letters were sent to addresses around the country and the person or persons responsible have still not been brought to account. Some of the anthrax had been specially treated in order to make it very dangerous to humans if inhaled into the lungs. Fortunately, only a small number of people were killed, but there was widespread disruption and concern as a result of the attacks, and the authorities in many countries are clearly still worried that further attacks with biological weapons may occur in the future. A new threat has arisen for us all – bioterrorism – and a new policy concern also – prevention.

At the same time, there was a disastrous failure by the international community to reach an agreement to strengthen the key Biological and Toxin Weapons Convention (BTWC) with a verification protocol after almost ten years of discussion and negotiation. So, in my opinion, we face a looming threat against which we have an ailing control system. This book is not a value-free account of the problem. I believe that we must work to build a safer world through the steady development of international law – including internationally agreed measures of arms control and disarmament.

On the other hand, I have no wish to present a one-sided polemic, so I have tried to provide sufficient references in each chapter for my arguments to be checked out and alternative viewpoints investigated. However, in a popular account intended to be read by non-specialists, I have avoided referencing every point as I would do in a scientific paper. Anyone who wishes to investigate the issue in depth will have no trouble finding a mass of information on the internet. Good starting points are the websites we have developed at Bradford[1] and the quarterly bulletin of the Harvard–Sussex programme.[2]

My intention has been to produce a reasonably comprehensive and self-contained introduction to this complex subject which could be of interest both to general readers and to students in biology or international security courses. To this end I have provided a good number of cross-references between chapters and for a small number of critical points I have also reiterated the information so that the argument is clear. Many people still do not grasp how advanced the offensive biological weapons programmes of the last century had become. I have therefore introduced a number of figures in some chapters in order to illustrate this point. These figures come preferentially from US sources, since much (carefully edited) material was previously released under the Freedom of Information Act there. The use of this US-sourced material should not be taken to indicate that the US was the only country with a biological weapons programme in the last century, merely that more material is available from that source. For ease of reference I have also included a good number of summary tables, for example giving lists of dangerous pathogens that might be misused for hostile purposes and the outlines of international agreements of relevance.

I would like to thank the many friends and colleagues whose written work, presentations and discussions at various meetings since I began researching biological disarmament in earnest after the first Gulf war in 1991 have been of great help to me when writing this book. I should also like to thank the British railway system for their inadvertent assistance. I have spent the autumn of 2004 and spring of 2005 touring the UK to discuss with scientists what might be done to avoid the misuse of their work (see Chapter 8) and this has allowed many hours of writing en route. I am sure the writing of this book would have been much more difficult without those hours away from the other distractions of work in a university department.

I thank the colleagues who have read and commented on my book, in whole or part, for eliminating some of the inevitable errors of fact and miscommunication in the original draft. I am also indebted to a very careful referee who read the whole text for my publishers. Any errors that remain in the book are mine. Finally, I would like to thank my wife Janet for all the help she has given both as a fellow biologist and as a general reader.

## references

1. See <www.brad.ac.uk/acad/sbtwc>.
2. See <www.sussex.ac.uk/Units/spru/hsp>.

Malcolm Dando

chapter one

# bioterror: threat and response

## introduction

Many politicians stated, and many people believed, that the world had changed after "9/11" (September 11, 2001). The threat from terrorists intent on causing mass casualties, perhaps even through the use of weapons of mass destruction – nuclear, chemical or biological – implied, it was said, the necessity of a radically new set of foreign and domestic policies. But how was the general public to understand the extent to which civilized society was now threatened? And how were they to properly judge the new policies that were being proposed and rapidly implemented? In particular, how were people to understand a complex issue such as biological warfare and biological terrorism when so little sound information was in the easily available public record?

Those with long memories will recall the flurry of interest in such issues in the late 1960s when President Nixon publicly abandoned the huge US offensive biological weapons programme and the international community moved to totally prohibit such weapons in the Biological and Toxin Weapons Convention (BTWC) which was agreed in 1972 and entered into force in 1975. Those with access to more specialized journals will also know that there has been an increasing number of articles on new biological weapons threats since the 1991 Gulf war which raised the spectre of possible Iraqi military use of biological weapons and since the true scale of the illegal Soviet offensive biological weapons programme became clear through evidence brought to the West by a series of defectors. Yet, for most members of the public, the information on biological weapons, biological warfare and biological terrorism that surfaces in the media is infrequent and fragmentary, making it difficult to form an integrated realistic picture of what is at stake.

The aim of this book, therefore, is to provide a straightforward and reasonably comprehensive account of what biological weapons are, how they have increasingly come to be seen as a threat as our understanding of the life sciences has developed over the last century, and what might be done now to prevent them becoming a terrible threat to human rights and human dignity in the coming decades.

## an unprecedented world?

Speaking to the American public in 1963, President Kennedy admitted that he was haunted by the prospect of fifteen to twenty states possessing nuclear weapons within a decade. Fortunately, a non-proliferation regime consisting of an integrated web of policies centred on the Nuclear Non-Proliferation Treaty has kept the number of nuclear-weapons states below ten to this day. One of the major reasons why this has been possible is the difficulty a potential proliferator has in producing the fissile uranium or plutonium material required.

Unfortunately, no such barrier exists for biological weapons. Obtaining biological and toxin agents, producing large quantities and even effectively weaponizing them, though not straightforward, have been shown to be perfectly possible in a series of offensive biological weapons programmes carried out by major states over the last hundred years.[1] Furthermore, the spread of new biotechnology capabilities will inevitably make such tasks much simpler for states or sub-state groups to carry out in the future. Such capabilities will continue to develop and spread because the ongoing revolution in the life sciences undoubtedly offers many potential benefits for human society, for example in healthcare and agricultural production. The problem is that the science and technologies developed for benign purposes can also be misused. If enough is known about the mechanism by which an infectious disease kills to prevent it killing, enough is probably also known about how to increase its virulence. This is a quite unprecedented problem. The question is "How do we regulate the revolution in the life sciences so that we encourage the growth of beneficial work while also preventing misuse?"

## international law

A moment's reflection will confirm that the misuse of modern biology cannot be prevented by national action alone. If a contagious agent like smallpox were to be used by terrorists to attack a country on the European mainland, only co-ordinated international action would be adequate to neutralize the great danger posed to many other countries inside and outside Europe. It may not be widely appreciated, but as the modern world has become more complex and interrelated since the Second World War, there has been a concomitant major growth in international law. Reflecting the conditions of modern international society, it has been necessary to develop law to govern numerous new areas of activity such as the air and outer space, and new uses of familiar areas such as the sea. International law has also developed to govern human rights, the use of force and how treaties are made.

Not surprisingly, therefore, the development of modern biology and its industrial use has also seen the growth of international agreements for its regulation. These agreements may be classified in three groups: health, disease and development; trade and environment; and protection against misuse.[2] The work of the World Health Organization and the Food and Agriculture Organization comes into the first category, and that of the World Trade Organization and World Intellectual Property Organization, as far as they cover biotechnology, comes into the second category. In regard to protection against misuse, the UN Drug Conventions and the World Anti-Doping Code, for example, would be relevant. So too would be the 1925 Geneva Protocol which bans the use of chemical and biological weapons, the 1975 BTWC which prohibits the development, production, stockpiling or other acquisition of such weapons, and the 1997 Chemical Weapons Convention (CWC).

## arms control and disarmament

Like other international legal agreements the Geneva Protocol, the BTWC and the CWC were a reflection of their times. Following the Second World War there was an attempt to achieve "General and Complete Disarmament" but, given the suspicions of the times, this failed. During the 1960s a more limited concept of arms control arose which tried to prevent use of the newly developed, devastating, nuclear weapons by clearly marking the difference between them and conventional weapons, by creating stability through both superpowers having survivable strategic nuclear systems, and by preventing nuclear proliferation. On this basis, a series of agreements were reached, both bilaterally between the superpowers and multilaterally by the international community. By the end of the cold war period, Jozef Goldblat, in his magisterial survey, *Arms Control: A Guide to Negotiations and Agreements*,[3] was able to list a huge range of areas where agreements had been reached both before and during the cold war:

- nuclear weapons explosions
- nuclear arms limitations
- nuclear weapons proliferation
- chemical and biological weapons
- environmental and radiological weapons
- outer space and celestial bodies
- the sea environment
- demilitarized areas
- denuclearized zones
- confidence building in Europe
- reduction of forces in Europe
- constraints on conventional arms transfers
- restrictions on the use of weapons
- prevention of accidental war

Against that background, though it was clearly not going to be easy to achieve, the idea of a peaceful world in which the level of armaments was minimal again appeared a possibility to some. The agreement to extend the central Nuclear Non-Proliferation Treaty indefinitely in 1995 was accompanied by unanimous adoption of a new set of principles and objectives for nuclear non-proliferation and disarmament. This reaffirmed that the ultimate goal of the treaty was the complete removal of nuclear weapons and disarmament under international control. The roadmap of agreements needed in the meantime seemed clear – a Comprehensive Nuclear-Test-Ban Treaty, a convention banning the production of fissile material and so on. On the back of such positive developments Frank Blackaby, a former director of the prestigious Stockholm International Peace Research Institute, argued that a Nuclear Weapons Convention embodying a prohibition equivalent to that in the BTWC and CWC might be envisaged.[4]

Blackaby argued realistically that the acknowledged nuclear-weapon states might find these ideas attractive because failure to capitalize on the opportunity presented by the end of the cold war through further multilateral agreement was likely to leave them

> far more vulnerable in the long run to new, highly destabilising and inherently unmanageable security threats, *pre-eminent among them proliferation of weapons and materials of mass destruction.* (Emphasis added)

We now know, of course, that the multilateral road was not taken. The Comprehensive Test Ban Treaty was duly agreed in 1996 but was limited by its entry-into-force requirements and was totally rejected by the US Senate in 1999. Significantly, India and Pakistan then joined the nuclear club with a vengeance in their series of 1998 nuclear tests. Yet, as the 2000 US presidential election approached, all did not seem lost, for the Democrats still espoused the rationale of multilateralism and sought to retain and develop what had been achieved.

## a different model

The Democrats, however, did not assume power after the election, but the neo-conservative George W. Bush administration did. The new leaders believed that the United States should primarily use its vast "unilateral" military power to achieve its goals. They believed that the multilateral agreements of the past were merely entangling the US in rules that restricted its freedom of action. There followed rejection of the Kyoto agreement on climate change, the International Criminal Court and so on. In the area of arms control, Nobel Laureate Professor Joseph Rotblat, of the Pugwash movement of scientists, has shown where this approach is likely to lead. He noted a series of crucial policy documents:

- Nuclear Posture Review (January 2002)
- The National Security Strategy of the United States of America (September 2002)
- National Strategy to Combat Weapons of Mass Destruction (December 2002)
- National Policy on Ballistic Missile Defence (May 2003)

Then he stated[5]:

> These policies seem to have two aims: one, a defensive strategy to make the US invulnerable to attack from outside; the second, an offensive strategy, to threaten an unfriendly regime with military action, including the use of nuclear weapons, if it attempts to acquire WMDs [weapons of mass destruction] for itself.

In his opinion this "might is right" option cannot last:

> Even if the Americans were less arrogant in pursuing that role than they are now, *a system with a built-in inequality is bound to be unstable.* It is bound to create resentment, a resentment that will find expression in various ways, including an increase in international terrorism. (Emphasis added)

It could clearly be argued, then, that the world security situation did not change on September 11, 2001 but instead changed when a new administration with a radical world view took office in Washington in January 2001. The vicious attack on September 11 was therefore viewed by some in Washington as an attempt to exploit perceived US weakness and required a response that demonstrated the new US strength and determination. At the time of writing, the consequent overestimation of American ability to rule the world is all too obvious in the increasing quagmire that is Iraq.

## controlling biological weapons

The difficulties of US policy in Iraq are in the media headlines and plain for all to see. Much less known is the failure of policy in regard to biological warfare and terrorism. As part of the general move to develop arms control in the early 1990s, the states party to the Biological and Toxin Weapons Convention agreed in 1991 to try to remedy its major deficiency. As an agreement of the middle of the cold war era the BTWC was significant in the sweeping nature of its prohibition but deficient in its lack of effective verification provisions. There was no means of checking that the parties were living up to their obligations – as was amply demonstrated by one of the three depositary states, the Soviet Union, carrying out a huge illegal offensive biological weapons programme. As we shall see, by mid-2001 there was great hope that a verification protocol developed by a decade of difficult negotiations could be agreed, but the new

US administration rejected both the proposed protocol and the mandate on which it had been negotiated. Thus, in an area of real concern over proliferation because of the dual-use nature of the technology, a crucial element in the web of policies required to prevent proliferation of biological weapons capabilities has been left extremely weak.

As we shall see in later chapters, a vast amount of money has been invested in new biodefence research in the United States. Since this research is concentrated in areas of direct concern for biowarfare and bioterrorism – for example, how crucial pathogens cause disease – it is bound to add greatly to knowledge that could be misused. Worst of all, some of this research could easily be perceived by other countries as being offensively rather than defensively orientated. The history of such misperceptions means the danger of responses in kind is far from unlikely.

This book is based on the proposition that there is no alternative to a joint co-operative international response to the problem of controlling biowarfare and bioterrorism. Though individual, sub-state, national and regional policies will be required in an integrated web of preventive policies, at the core we need the international disarmament prohibition embodied in the BTWC. Furthermore, this convention will have to be strengthened eventually as part of a renewed multilateralism if we are to succeed in preventing a biological arms race. At the time of writing, it is unclear whether US security policy will change following the 2004 presidential election and whether there will be a reassessment in Bush's second term. The most likely prospect, however, is that there will be no substantial change until 2008. A period of stagnation for the control mechanism and very rapid developments in the life sciences seem the most likely prospects for the near future.

## outline of the book

Bearing all that in mind, Chapter 2 gives an account of the largely unknown history of biological weapons and offensive biological weapons programmes before 1945. There were probably instances of biological weapons use before the late nineteenth century but such attacks were not carried out on the basis of valid scientific knowledge. Only with the work of scientists like Pasteur in France and Koch in Germany was a proper understanding possible of the bacterial agents that cause diseases such as anthrax and glanders, and then both of these agents were used to attack valuable draught animal stocks in the First World War. The chapter then reviews the offensive programmes of the interwar years, particularly the Japanese programme and its attempts to use biological agents against people in China, and afterwards the programmes of the Second World War, for example of the United Kingdom.

Drawing on recent studies, Chapter 3 describes the offensive programmes that followed the Second World War through to the conclusion of the BTWC. The descriptions of the programmes in the United States, the United Kingdom, Canada and France make it clear that at the beginning biological weapons were regarded as just as threatening as weapons of mass destruction as were the new nuclear weapons. Only when nuclear weapons became more readily available were biological weapons downgraded, and then almost certainly *not* because they were of little use but rather because they might be more easily obtained by states with limited technical and financial resources.

Chapter 4 deals with the period from the agreement of the BTWC through to the present day. There are probably a number of other states that have or have had offensive biological weapons programmes during that period, but the three we know about were in Iraq, South Africa and the former Soviet Union. The huge programme initiated in the USSR after the agreement of the BTWC is the most significant for the future, since it clearly involved the first application of the new biology to making more effective biological weapons. The exact details of what was done remain obscure, but enough is known to provide a clear warning to us all.

Chapter 5 reviews the nature of biological agents. It deals with anti-personnel biological agents, firstly the most dangerous of these (Category A agents) and then the less dangerous (Category B and C agents). A brief description is given of each of the major agents and the reasons they are of such concern. The chapter also covers the often neglected, but very important, issue of anti-agriculture biological attacks and finishes by reviewing agent production and dissemination technologies.

The impact of the biotechnology revolution is introduced in Chapter 6 through a review of what the states party to the BTWC concluded at their successive five-yearly review conferences in 1980, 1986, 1991, 1996 and 2001–2. The description reveals the growing appreciation of and apprehension about the impact of the developing science and technology on the potential for biological warfare.

Such apprehensions have greatly increased within the scientific community because of a series of experiments in the civil sector since 2001 which have elevated the perceived risk of possible misuse of benign research. Some of these experiments are described.

With a firm historical and scientific understanding in place, Chapter 7 deals with the vexed question of the real nature of the threat we face today and in coming decades. The chapter reviews anti-personnel WMD and non-WMD attack scenarios and anti-agriculture possibilities. It is suggested that WMD attacks on people by terrorists are still very unlikely, but that non-WMD attacks on people and severe attacks on agriculture are technically possible. Moreover, catastrophic terrorism is something we certainly cannot rule out for the future.

Chapter 8 discusses how the misuse of biology can best be prevented. Taking up the idea of a wide-ranging web of preventative policies that together suggest to potential proliferators that it is not worth developing biological weapons, the relevant policies – from intelligence to export controls, arms control, detection and protection, and an international willingness to act against violators – are reviewed in turn. The chapter ends by cautioning against the inadvertent generation of action/reaction cycles of biodefence and offence, since this could well happen if the vast increases in biodefence spending are not clearly understood to be solely for defensive purposes.

Chapter 9 covers the evolution of the prohibition of biological weapons in international law from the 1925 Geneva Protocol through the 1975 BTWC and the 1997 Chemical Weapons Convention. Particular attention is given to the structure and deficiencies of the BTWC and the long series of efforts to remedy these deficiencies. Following the failure of the verification protocol negotiations in 2001 a more limited set of objectives was agreed for the years leading up to the sixth review in 2006. This new inter-sessional process is discussed and the prospects for 2006 and beyond are set out. The conclusion is that there is much to be done to improve this patchwork quilt of controls in the years ahead.

The concluding Chapter 10 contrasts two possible futures: one in which an action/reaction, offensive/defensive arms race leads to the development and production of ever more advanced – and dangerous – biological agents; and another in which we succeed in the steady development of ever more effective constraints on the malign misuse of biology, one in which the huge potential benefits of the new biology become increasingly available to all.

## references

1. Wheelis, M., Rozsa, L., and Dando, M.R. (eds) *Deadly Cultures: Bioweapons from 1945 to the Present.* Cambridge, MA: Harvard University Press (in press).
2. See <www.brad.ac.uk/acad/sbtwc/gateway/>.
3. Goldblat, J. (2002) *Arms Control: A Guide to Negotiations and Agreements*, 2nd edn. London: Sage.
4. Howard, S. (2003) A receding disarmament horizon? Lessons from an era of retreat and defeat. *Disarmament Diplomacy*, October/November, 3–15.
5. Rotblat, J. (2003) The nuclear issue: Pugwash and the Bush policies. *Pugwash Newsletter*, December, 57–63.

chapter two

# biological warfare before 1945

## introduction

Many people will recall hearing stories that suggest the use of biological warfare in antiquity and the Middle Ages, like the Tartar attack on the well-defended seaport of Caffa (now Feodosia in the Ukraine) in 1346 when plague-infected corpses were catapulted into the city in an effort to break the resistance.[1] The retreating Genoese forces are thought to have then at least contributed to the subsequent outbreak of "Black Death" in Europe. Similarly, there are many suggestions, for example during the American Civil War, of animal carcasses being used to make water supplies unusable.

In a detailed analysis, Mark Wheelis argues that all these accounts need to be examined critically.[2] There is always a likelihood of increased outbreaks of natural disease in the disrupted social and economic conditions of warfare and the obvious possibility also of the opportunity being taken to make false accusations. Furthermore, the longer ago an incident is thought to have occurred, the less likely there is to be good credible documentation. Finally, since the agents and underlying mechanisms of infectious diseases were not properly understood until the latter part of the nineteenth century, any *effective* use of biological warfare or terrorism based on *scientific analysis* of what was required was virtually impossible. Wheelis therefore argues for the use of a very strict set of criteria to judge the credibility of accounts of pre-scientific biological warfare. He believes that there is probably a good deal of historical documentation in languages other than English, particularly relating to South America and Asia, that remains to be seriously analysed. However, at least one historical example for which there is evidence in English stands up against his very strict set of criteria, set out below:

- It should make political and historical sense.
- It should be supported with sufficient detail to allow evaluation.
- The alleged action should be technically feasible in the context of the state of knowledge at the time.
- The reported outbreak should be a plausible consequence of the alleged action.
- The source of the allegation should be clearly documented.
- There should be some evidence to support it.

This convincing example concerns the deliberate use of smallpox by the British against North American Indians at Fort Pitt in 1763. Following the French surrender of Canada, Pontiac, a major chief of the Ottawa Indians, united tribes along the frontier from Virginia to New York and attempted to drive the British out of Canada and the Mississippi watershed, with the intention of returning the territories to the French. The Indians resented both the fact that the British had broken promises made to them during the war and that the British were much more restrictive in trading powder and shot with the Indians than were the French.

Many of the British troops had returned home after the defeat of the French and those remaining were badly overextended when simultaneous Indian attacks were made. Eight forts were overrun and settlers were killed and captured in large numbers. Survivors made their way east to larger settlements and large-scale loss of territory appeared inevitable. Even Fort Pitt, a major outpost, seemed likely to fall. The Indians wished to persuade the British to leave without a fight and a meeting was held between the besieging Indians and representatives of the fort garrison. The British contingent included a Captain Ecuyer, commanding officer of the fort, and William Trent, a trader who was commanding the militia (but reporting to Ecuyer). The British refused to leave the fort, even though smallpox had just broken out in the crowded conditions there.

Trent's journal for 24 June 1763, the day of the meeting with the Indians, gives an account of the meeting which reads, in part,

"Out of our regard to them, we gave them two Blankets and a Handkerchief out of the Small Pox Hospital. I hope it will have the desired effect." This transfer of infectious material is confirmed by an entry in Captain Ecuyer's ledgers which lists obligations to Trent's company, including:

> To Sundaries got to replace in kind those which were taken from people in the Hospital to Convey the Smallpox to the Indians viz;
> 2 Blankets..........................................at 20/ £2: 0: 0
> 1 Silk Handkerchief 10/ &1 Linnen do: 3/6 0:13:6

The account was signed by the Captain, who added, "I do hereby certify that the above articles were had for the uses above mentioned." The account was countersigned by the officer who dealt directly with transactions between Trent's trading company and the Crown and by General Gage, the British commander-in-chief of forces in America.

Though there *was* an outbreak of smallpox among the relevant Indians about the time of the above incident, it is unclear whether the action of the British at Fort Pitt was the cause. On the other hand, documentation exists of discussions at the time between other high-ranking British officers of using precisely this method – giving infectious blankets to the Indians in order to cause smallpox – and this is clearly what was intended by those who carried out the biological attack at Fort Pitt. Though smallpox was a dreadful disease for the British, their population had much more experience of it and was thus much less susceptible to it than the Indians. It is also likely that the British attempted to use smallpox against the Continental Army during the American Revolutionary War. Looking at the historical record as a whole, Wheelis comments,

> it is surprising that biological attack has not been more common than the record suggests. *Most likely, it was restrained by utilitarian difficulty and hazard* when cultural norms against it were weakened by exigency and self-interest. (Emphasis added)

However, the revolution in biology which took place in the last two decades of the nineteenth century, a period commonly called "the Golden Age of Bacteriology", was to initiate the scientific and technological understanding of microbial pathogens which would progressively remove such "utilitarian difficulty and hazard".

## microbiology in the late nineteenth century

At the beginning of the nineteenth century there were still many theories about the causes of disease which seem very strange to us today. These included supernatural forces, noxious vapours and imbalances in the four so-called humours of the body.[3,4] The changes that led to our present understanding did not come suddenly at the end of the century but were preceded by a steady accumulation of knowledge and better practice. Joseph Lister, for example, developed his system of antiseptic surgery upon the supposition that it would prevent micro-organisms from entering wounds. Nevertheless, two scientists, Louis Pasteur in France and Robert Koch in Germany, are generally regarded as having been particularly significant in developing the theory that microbial agents (germs) cause infectious disease.

One idea widely held at the time was that life could generate spontaneously. People believed, for example, that maggots arose in that way from decaying meat. Pasteur carried out a classic series of experiments in which broth was put into flasks and heated. The flasks, however, had long U-shaped 'swan' necks so that dust particles and microbes settled in the bottom of the U-bend of the neck despite the fact that the broth in the flask was open to the air. As we would now expect, the broth remained sterile and the idea of spontaneous generation had to be rejected. One reason for Pasteur's interest in this issue was that he was keen to show that fermentation of wine was caused by micro-organisms and that proper selection of the micro-organisms would influence the quality of the wine. He also showed that heating the final product of a fermentation for a few minutes at 50–60°C preserved it by destroying any micro-organisms present. We know this process today as "pasteurization" and use it extensively in the food and drinks industry.

Pasteur went on to play a major role in the development of the germ theory of disease and in the development of methods of vaccination using attenuated (weakened) strains of bacteria and against rabies virus as understanding of these agents increased. It was Robert Koch, however, who scientifically demonstrated how bacteria cause disease, through his work on anthrax. Koch injected healthy mice with material from diseased mice and the healthy mice became ill. He did this successively through a series of mice and then incubated a piece of infected spleen containing the anthrax bacillus from the last mouse in the series in a beef broth. As we would now expect, the bacilli grew, reproduced and in the right conditions also produced the environmentally resistant spores that the bacillus forms as part of its natural life cycle. When either the bacilli or the spores were injected into healthy mice, anthrax disease developed. Koch had therefore shown that anthrax germs caused the observed disease.

Koch put forward his famous postulates which had to be met if such a causal relationship were to be proved. In summary, these are:

1. The micro-organism must be present in every case of the disease but absent from healthy organisms.
2. The suspected micro-organism must be isolated and grown in pure culture.
3. The disease must result when the isolated micro-organism is inoculated into a healthy host.
4. The same micro-organism must be isolated again from the diseased host.

Application of this kind of scientific methodology then led to the remarkably rapid elucidation of the causes of many of the major bacterial diseases. Some of these successes of the Golden Age of bacteriology are shown in Table 2.1. It is very difficult for us, at the beginning of the twenty-first century, to imagine what it was like to live before this magnificent scientific achievement and its application in disease prevention, diagnosis and treatment. In particular, infant mortality rates dropped spectacularly, which led eventually to the reduction of birth rates to their present low level in the developed world. Unfortunately, the knowledge could also be used for hostile purposes, as was well illustrated in the First World War.

## biological warfare in the first world war

The ancient taboo against the use of poison in warfare had begun to be codified more clearly in the Brussels Declaration of 1874 and the two Hague Conventions of 1899 and 1907, but it appears that the German General Staff took the prohibition to be limited to attacks on humans. During the First World War draught animals were of crucial importance and many of the bacterial diseases, by then well understood, were natural pathogens of such animals. Germany

**Table 2.1** Discovery of major pathogens*

| Disease | Agent | Discoverer | Date |
|---|---|---|---|
| Anthrax | *Bacillus anthracis* | Koch | 1876 |
| Typhoid fever | *Salmonella typhi* | Eberth | 1880 |
| Glanders | *Pseudomonas mallei* | Loeffler & Schütz | 1882 |
| Cholera | *Vibrio cholerae* | Koch | 1883 |
| Undulant fever | *Brucella* spp. | Bruce | 1887 |
| Gas gangrene | *Clostridium perfringens* | Welch & Nuttal | 1892 |
| Plague | *Yersinia pestis* | Kitasato & Yersin | 1894 |
| Botulism | *Clostridium botulinum* | Von Ermengem | 1896 |
| Whooping cough | *Bordetella pertussis* | Bordet & Gengou | 1906 |
| Rocky Mountain spotted fever | *Rickettsia rickettsii* | Rickets | 1909 |
| Tularemia | *Francisella tularensis* | McCoy & Chapin | 1912 |

* From Prescott et al., op. cit.; Pelczar et al., op. cit.

certainly included the use of biological agents to damage draught animal stocks in its sabotage operations against the Allies, and there is more limited evidence also that France used the same methods against the Germans.

Germany was subject to a British naval blockade, and therefore the productive resources of the United States could only be used by Britain and its allies. Though the US did not enter the war until 1917, Germany carried out an effective sabotage campaign in the United States well before that in an effort to disrupt supplies to the Allies. As part of that campaign, efforts were made from 1915 through until the autumn of 1916 to use biological agents to infect horses being readied for shipment on the eastern seaboard. The key agent in this operation was Anton Dilger. His parents had emigrated to the United States from Germany and, although born in the US, Dilger returned to Germany and was educated there. However, Dilger, a physician in his early thirties at the time of the war, had maintained his US passport. He was therefore free to return to America in April 1915. Dilger's brother later testified:

> My brother was a doctor, and his part of the work had principally to do with making disease cultures for Germany to use in inoculating

animals with disease, [animals] that were being purchased in the United States for the Allied Governments.

Dilger's brother helped him to grow cultures of *Bacillus anthracis* (anthrax) and *Pseudomonas mallei* (glanders) in the basement of a house in Washington. These agents were then, for example, painted onto the nostrils of the purchased horses by a group of people recruited by the master of a German steamship trapped in the United States by the British blockade. From the evidence available, it is not clear how effective this part of the sabotage campaign was in reducing the supply of horses. Dilger left for Germany in January 1916 and attempted to return to the US in July 1917, but by then he was under suspicion and left quickly for Mexico. He apparently died in Spain of the "Spanish flu" in October 1918, after being awarded the Iron Cross Second Class in January 1918.

Germany carried out similar sabotage operations in neutral Romania until that country joined the Allied side in August 1916 and the operations became impossible. The main difference there was that it was possible for the cultures to be grown in Germany and shipped to Romania for use. Three shipments have been documented; two of these were probably used and the third was recovered by the Romanian authorities from the garden of the German Legation. The contents of the bottles found in the Legation were analysed at the time by the Institute of Pathology and Bacteriology in Bucharest and found to contain, in separate tubes, the causative agents of anthrax and glanders.

German sabotage operations were also carried out in Norway and Argentina. The Norway operations are of interest because the agents were in small capillary tubes embedded in sugar cubes – presumably to be fed to the animals. One of these sugar cubes, confiscated by the Norwegian authorities when the German operatives were expelled in 1917, was recently discovered in a police archive and analysed by modern methods (polymerase chain reaction, PCR). It was shown to contain *Bacillus anthracis*. The operations in Argentina are of interest because the material required had to be transported there. U-boats were used to ship the cultures from the then Austrian port of Pola on the Adriatic (now Pula, Croatia) to Spain and then they were taken on to Argentina by agents travelling on commercial steamships. These operations required the sending of secret wireless telegraph transmissions which, unknown to the Germans, the British could intercept and decode. Some forty transmissions dealing with biological sabotage were so intercepted and analysed by the British at the time. However, the British do not appear to have viewed the operations in Argentina as effective; although having considered biological warfare themselves, they did not at that stage think it feasible.

What can be deduced from the available information is that a large-scale effort was made by the German General Staff to use biological warfare agents in several different countries for a very serious military sabotage campaign. Soon after the Golden Age of bacteriology had elucidated the nature of infectious bacterial diseases, therefore, that knowledge was applied in deadly earnest in all-out warfare.

## biological warfare between the wars

The extensive use of chemical warfare during the First World War had horrified civilized society and it is therefore understandable that, in the aftermath of the war, efforts were made to limit the possibility of such use of chemical warfare agents in the future. These efforts eventually led to the agreement of the 1925 Geneva Protocol, now considered to be part of customary international law binding on all states.[5]

The Protocol has an initial statement that provides a link to the earlier nineteenth- and also twentieth-century agreements:

> Whereas the use in war of asphyxiating, poisonous or other gases, and of all analogous liquids, materials or devices, has been justly condemned by the general opinion of the civilized world; and
>
> Whereas the prohibition of such use has been declared in Treaties to which the majority of Powers of the World are Parties; and
>
> To the end that this prohibition shall be universally accepted as part of International Law, binding alike the conscience and the practice of nations.

Following this statement it was intended that there should be a "declaration" relating to chemical weapons. However, the Polish delegation pointed out that bacteriological weapons could be just as dangerous, so the declaration eventually agreed was:

> That the High Contracting Parties, so far as they are not already Parties to Treaties prohibiting such use, accept this prohibition,

> *agree to extend this prohibition to the use of bacteriological methods of warfare* and agree to be bound as between themselves according to the terms of this declaration. (Emphasis added)

It will be noted that the 1925 Geneva Protocol prohibits only the *use* of chemical and biological warfare. The parties to the Protocol were therefore free to continue developing and stockpiling such weaponry and many entered reservations to their agreement, making it clear that though they would not use such weapons first, they were prepared to use them in retaliation if these weapons were used first against them. In the event, a number of offensive biological weapons programmes were undertaken by major states in the years between the two world wars. It was not until the latter part of the twentieth century that further arms control restrictions were placed on chemical and biological weapons, extending the prohibition beyond just use and rendering the reservations to the 1925 Geneva Protocol invalid.

In France, despite the destruction of many of the relevant records in June 1940 (to prevent them falling into German hands), enough documentation remains to show that there was a very serious inter-war biological weapons programme. This had three phases: a period of intense interest following the First World War up until 1926; a dormant phase from 1927 to 1934; and then a marked increase in military activities until the fall of France in 1940. These phases clearly related to the fluctuation in France's relationship with Germany and to the signing of the 1925 Geneva Protocol, for which France is the depositary state.

A crucial 1922 French report, "The Use of Bacteriological Weapons in War", began by reviewing what was known of German activities, and it is clear that perceptions of what Germany might be doing were very important throughout the period of France's programme. The second part of the report discussed the possible advantages of using these weapons in war and concluded that, though not usable on the frontline, "they would be appropriate ... against such targets as civilian populations, urban centres, troop assembly points, barracks, stations, factories or industrial sites". The report also noted the possible use of these agents to contaminate food and water supplies or to attack draught animal stocks such as horses. Another section of the report dealt with experimental work that had been carried out to show how a cloud of microbiological agents could be created by an explosive device such as a bomb. After this report, a period of experimentation followed which led to full-scale trials of an experimental bomb dropped from an aircraft in 1926.

When tensions escalated in Europe in 1933–4, France refused to support German rearmament and stated that it would ensure its own security. As part of this process, the dormant biological weapons programme was reactivated and, again, concerns about possible German activities (received in intelligence reports) were clearly important. For example, reports that Germany had conducted trials in London to test the spreading of bacilli through the Underground network were taken seriously and led to tests on the Paris metro. These reports of German activities were incorrect, but the tests in Paris showed that bacilli could be spread in the underground network. It is certain that in the months before the German occupation progress was being made in the testing of operational weapons.

Data on most European countries have not been researched as they have for France, but it is known that there was an offensive biological weapons programme in Hungary between 1936 and 1944.[6] Located in Budapest, the programme is known about from a surviving report provided by the leader of the programme, Colonel Bartos, to the Ministry of Defence in 1955 when he returned from imprisonment in the USSR as a prisoner of war. Bartos' group considered three methods of attack with biological warfare agents: delivery by aeroplane or artillery; the use of saboteurs to spread the germs; and contamination of evacuated territory. They experimented to discover the optimal meteorological conditions for attack, the effective concentration of germs needed and the most suitable bacteria to use, such as *Bacillus anthracis* (anthrax), *Clostridium perfringens*, *Salmonella paratyphi* and *Shigella dysenteriae* (see Chapter 5). They also looked at enhancing the virulence of *Salmonella paratyphi* by passage through a series of animals and at how best to store the pathogenic bacterial strains for long periods. The Hungarians tried to develop connections with other programmes in like-minded countries. As we know now, there was little activity in Germany because of Hitler's opposition, but it appears that there was a programme in Italy and that exchanges of information and visits took place.

Outside Europe a huge offensive biological warfare programme began in Japan in the inter-war years (Figure 2.1). The driving force behind this programme appears to have been Ishii Shiro. As a young army medical doctor, Ishii concluded that the 1925 Geneva Protocol had a very particular meaning: "[h]e reasoned that if the weapons

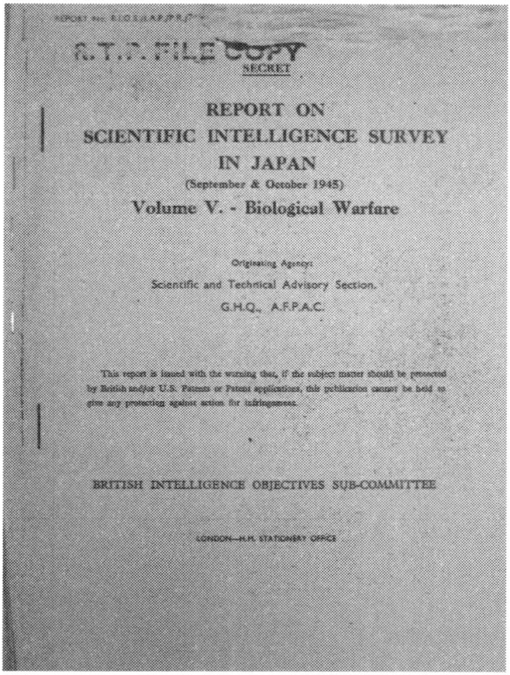

SECRET

REPORT ON
SCIENTIFIC INTELLIGENCE SURVEY
IN JAPAN
(September & October 1945)
Volume V. - Biological Warfare

Originating Agency:
Scientific and Technical Advisory Section.
G.H.Q., A.F.P.A.C.

This report is issued with the warning that, if the subject matter should be protected by British and/or U.S. Patents or Patent applications, this publication cannot be held to give any protection against action for infringement.

BRITISH INTELLIGENCE OBJECTIVES SUB-COMMITTEE

LONDON—H.M. STATIONERY OFFICE

**Figure 2.1** *Front cover of a UK report on Japanese biological warfare. This detailed British scientific report was prepared immediately after the Second World War and concluded in part that "Information has been obtained that from 1936 to 1945 the Japanese Army fostered offensive BW, probably on a large scale."*

were placed on a forbidden list, then Japan should possess them in order to acquire an advantage over its opponents in future wars". Furthermore, he argued that biological weapons had unique advantages for Japan because they did not require raw materials such as iron which Japan had difficulty in obtaining, and that, since biological weapons were self-multiplying in the victim, the quantities required were small.

Ishii's initial attempts to persuade his superiors were unsuccessful, but after he was appointed Professor of Immunology at the Tokyo Army Medical School in 1930 he began to achieve his aims. The Japanese offensive biological warfare programme ran from 1931 right through the Second World War to 1945. At its peak, the programme probably involved fifteen thousand people. It was unique in that it used human experimentation to test biological agents, which probably killed thousands of Chinese people, and it also employed biological agents in military field operations in China, which led to the deaths of many thousands more.

Ishii carefully distinguished between two types of bacteriological warfare research. Type B was orientated towards defence and concerned matters such as the production of vaccines and water purification. Type A, however, was offensive "assault" research. He argued that Type A research had to be done outside Japan (because it involved human experimentation) and he seized the chance to pursue such research when he was posted to Japanese-occupied Manchuria in China in 1932. There, some one hundred kilometres south of Harbin, he built a state-of-the-art "biological warfare death factory, the Zhong Ma Prison Camp", where the programme's efforts were eventually concentrated on anthrax, glanders and plague. Sometime in 1934–5 a prison rebellion led to the escape of some prisoners and a few made it to a communist guerrilla camp. Less than a year later a huge explosion rocked the Zhong Ma camp. The secret facility had presumably been revealed by those who escaped, and Ishii and his gang moved on to new sites.

The largest of the new sites was the infamous Ping Fan, located just twenty-four kilometres south of Harbin. When completed, this complex consisted of over 150 buildings, including laboratories, barracks, a large farm supplying the camp, and two special prisons – one for men and one for men, women and even a few children. The heart of the camp was the huge concrete and brick administrative building at whose centre were the two prisons. This arrangement made it virtually impossible to see the prisons from the outside. Underground tunnels led from the administrative building to the other facilities in the complex. There was an airfield near the camp and Ishii had a fleet of aircraft at his disposal. The camp also had a railway line linking it to Harbin. Enormous resources were poured into the offensive biological warfare programme, other major parts of which were run by Kitano Masaji at another site at Mukden from 1932 to 1942 (when he took over at Ping Fan) and Wakamatsu Yujiro at Mokotan, a suburb of Changchun. Wakamatsu's group (designated Unit 100) was mainly concerned with plant and animal research but was also involved in human experimentation.

It is particularly striking that Ishii and his colleagues were able to recruit many able scientists for this dreadful programme. One pharmacist told a postwar investigator, "We did not think that way [re issues of morality]. We were working for our country. We did as we were told. I thought General Ishii was a great man, an important man." Indeed, one of the people involved in the programme suggested that most microbiologists in Japan were connected in some way with Ishii's activities. Very few are known to have objected.

Many of the relevant records were destroyed at the end of the Second World War, so that today it is impossible to give accurate figures of the number of people killed in experiments. We know, however, that people were dosed with pathogens to determine the amount needed to kill, given drinks laced with germs to discover the best fluids to use, tied to stakes to test the effective range of biological bombs and so on. The victims usually survived for two to six weeks; some were dissected while still alive; others were dispatched by poison or a bullet to the head. In order to have the necessary material for testing, and for vaccine production, capabilities were developed for producing kilogram quantities of bacteria for plague, anthrax, typhoid, cholera, dysentery and other diseases.

Ishii and his colleagues were particularly interested in discovering the minimum dose necessary to infect fifty per cent of those who came into contact with the pathogens. This figure was determined for each pathogen by experiments on ten to twenty individuals. In the main, efforts were concentrated on bacterial agents and little effective work was done on viruses, rickettsiae or toxin agents. Curiously, despite its seemingly scientific approach, at least fifteen years of work produced no really *effective* means of delivering these biological agents via a bomb or shell. Nevertheless, by 1939 Ishii was ready to carry out large-scale field tests.

## biological warfare in the second world war

From May to September 1939 Japanese troops were engaged in fighting with Soviet troops on the Soviet–Manchurian and Manchurian–Inner-Mongolian borders. The Japanese army suffered a series of heavy defeats and in these circumstances Ishii was able to persuade his superiors to use his biological weapons capabilities. He used both personnel on the ground committing acts of sabotage and the firing of artillery shells filled with germs. Many troops on both sides suffered from plague, dysentery and cholera, but it is uncertain whether this was the result of Ishii's actions or because of the primitive sanitary conditions in the area. Nevertheless, Ishii convinced his superiors that his work was successful and the group – now designated by its infamous name, Unit 731 – was given a special commendation. Ishii himself received a decoration for his services to Japan.

From then till 1942 Japan carried out a series of large-scale biological weapons attacks in China. For example: "In 1940, Unit 731, in conjunction with other BW [biological warfare] units, caused a devastating cholera outbreak in Changchun, the capital of the puppet [of Japan] Manchukuo regime" and "in July 1942, Ishii and his confederates spread more than 130 kg of paratyphoid A and anthrax germs throughout a large area surrounding Nanking in central China. Several epidemics ravaged the region later that summer and autumn" and "In other tests, plague-infected rats were let loose by Unit 731 into densely populated communities. Plague epidemics erupted shortly afterwards. Ningbo in Zhejiang Province was hard hit with numerous epidemics as a result." After Ishii was transferred from Unit 731 in August 1942 there were fewer such field trials of biological agents. Ishii's successor Kitano preferred to concentrate on laboratory killings.

One further attempt to use biological weapons could have been made late in the war when US authorities were concerned that biological agents could have been one of the payloads used by the Japanese in their air balloon attacks on the United States. In these attacks the Japanese sent air balloons across the Pacific.[7] In winter a traverse was possible in about three days by riding the balloons on the jet stream at around ten thousand metres. Some 9300 balloon bombs were launched in the winter months of 1944–5. Although the attacks appear not to have involved biological agents, with a munitions-carrying capacity of thirty kilograms it was clearly theoretically possible to initiate damaging epidemics in agriculture. A final irony is that the Soviet Union put some of the Japanese biological weaponeers on trial in the late 1940s. This was dismissed as cold war propaganda in the West – where the United States had granted immunity to Ishii and his colleagues in exchange for their data.

Although Germany had ratified the 1925 Geneva Protocol in 1929, foreign governments became concerned when the Nazis rose to power and Germany left the League of Nations in 1933. Given the success that Germany had in rearming and the obvious ruthlessness of the Hitler regime, it is understandable that foreign intelligence agencies feared the worst in regard to German preparations for biological warfare. As we now know, they were wrong. Elements within the German military and government worried about what was being done in other countries, as when they had access to French material after 1940, and when they learned about possible Soviet activities from prisoners of war during the campaign on the Eastern Front, but Hitler is now known to have prohibited offensive biological warfare activities. The reasons for Hitler's decision are not known, but many of the German commanders and the scientific community supported the decision. What little activity occurred could hardly be called an offensive programme. This was therefore a classic case of misperception in which other countries reacted to what they thought Germany was doing when in fact Germany – though fearing the worst about other countries' activities – did not have an offensive programme.

Russia suffered thousands of casualties from chemical weapons attacks in the First World War, so it should be of no surprise that the leadership of the Soviet Union wished to be well prepared to deal with future chemical and biological weapons threats. During the 1920s secret joint chemical weapons tests were carried out in Russia with the Germans. However, German restrictions on the exchange of information and then Hitler's termination of the collaboration fuelled Soviet suspicions of Germany. From the 1920s onwards the Red Army began to investigate biological warfare, the scientist Jacov Fishman taking a leading role until he suffered in the purges of the mid-1930s.

The Soviet Military Chemical Agency (MCA) studied a range of agents in the laboratory and in subsequent field tests. These included *Bacillus anthracis* (anthrax), *Clostridium botulinum* (the bacterial source of botulinum toxin), *Yersinia pestis* (plague) and foot-and-mouth disease virus (for use against animals). However, during the purges of the mid-1930s large numbers of biological specialists were arrested and some of the accused were charged with biological sabotage. This purge of specialists must have had an effect on Soviet capabilities during the Second World War: although the Soviets certainly developed and used vaccines for defensive purposes, and there were reports of their interest in offensive activities, it seems unlikely that the Soviet Union really had an offensive capability *during* the war. A firm conclusion will only be possible when, if ever, the archives of the period become available to scholars.

The same is not true of Britain. By 1936 the Committee of Imperial Defence had established a Bacteriological Warfare Subcommittee. From 1940 this committee directed both defensive and offensive studies. The offensive activities were carried out in a much more systematic and scientific manner than those of the Japanese; they led to a usable anti-animal anthrax weapon and also cracked the main problem of using anthrax as an anti-personnel weapon.

The United Kingdom ratified the 1925 Geneva Protocol in 1930 but with the reservations that it would be bound only in regard to other parties to the Protocol and that the Protocol would not be binding with respect to states and their allies who broke the prohibition (in other words it reserved the right to retaliate in kind). The UK knew of German biological weapons activities during the First World War and officials were worried by intelligence reports and rumours – such as those about German tests in the London Underground in the early 1930s. All this eventually led to the formation of the Bacteriological Warfare Subcommittee. In November 1939 this was transformed into the War Cabinet Committee on Biological Warfare which, under the chairmanship of Lord Hankey, had offensive activities as part of its remit. Paul Fildes, an eminent civil scientist, was appointed to head the biology department at Porton Down to carry out the necessary work.

The immediate need in 1940 was for some means to retaliate in kind if Germany should use biological warfare. This need was met by producing five million cattle cakes laced with anthrax spores. These were to be dropped from planes onto German farming land in order to wipe out cattle whose loss would then deal an economic blow to Germany's overstretched agricultural system. Tests at Porton Down determined how many spores were needed to cause a lethal infection in cattle and the cattle cakes were designed so that there was no danger of aircrew being contaminated when they were used. Trials also showed that cattle would easily find the linseed oil cattle cakes if they were distributed sparsely on the ground. The cattle cakes were made by a London soap manufacturer and then sent to Porton Down, where a simple system allowed them to be charged with a suspension of anthrax spores and then sealed, dried and packaged in cardboard boxes. Each box contained four hundred of the cattle cakes. Production started fifteen weeks before Christmas 1942 and was complete by the end of April 1943. About twenty people were involved in the production, which was carried out without immunizations, safety cabinets or respirators.

If retaliation had been necessary, it was envisaged to involve either one huge 1250-aircraft raid or two or three smaller ones. Each aircraft would have had nine or ten boxes of cakes and these would have been disseminated over 18–22 minutes at two hundred miles per hour, the aircraft covering sixty miles of land. The aircraft had flare-chutes so a box could be emptied out manually every two minutes through the chute. In the event, the five million cakes were retained unused until the end of the war. The vast majority were then incinerated but one cake was found to still contain virulent spores when cultured in 1952. The last were destroyed in 1972.

Production of the anthrax spores was as simple as the distribution. At first, enamelled iron trays of about 15 × 10 × 2 inches containing solid agar-based medium were seeded with spores and the subsequent growth was collected with a hand-held suction device. The resulting mixture was milled with glass beads and filtered to give a suspension concentrate. Subsequently, fifty-litre milk churns set in a hot room and subject to periodic stirring were used. Later still, heating elements were put into the churns and agitation was achieved through an air-sparger. After sporulation was complete, the suspension was precipitated and concentrated.

Since he was based at Porton, the centre of British chemical weapons expertise, Fildes was able to draw on the experiences of the First World War in designing an anti-personnel biological weapon. He quickly decided that if an agent could be aerosolized and taken into the lungs to cause an infection an effective weapon might be possible. He decided to concentrate on two weapons agents: anthrax and, to a much lesser extent, botulinum toxin. Assisted by able scientists such as D.W.W. Henderson (subsequently director of the later Microbiological Research Establishment at Porton) and D.D. Woods (later Professor of Chemical Microbiology at Oxford), Fildes began work on an apparatus that produced bacterial clouds in which experimental animals could be infected and the required doses determined. It was also shown that a bursting munition could be used to create the required aerosol. For safety reasons, tests were then moved to Gruinard Island, off the Scottish coast.

On Gruinard in 1942 a combined team from the biology department and the chemical defence group at Porton worked with the military to test anthrax as a real weapon. A modified, thirty-pound, chemical high-explosive bomb was charged with a suspension of anthrax spores. The bomb was suspended from a fixed frame and sheep and air samplers were located in an arc 90–100 yards downwind of the device. On detonation, a cloud could be seen moving downwind. Seven days later, all but two of the sheep were dead. Later tests in 1942 showed that a lethal effect extended some 250 yards downwind. Calculations suggested that death was likely up to four hundred yards downwind and that the weapon was more potent than any known chemical weapon. Later, in October 1942, a trial was carried out in South Wales in which an anthrax-charged bomb was successfully launched from a bomber aircraft. Subsequent trials in Scotland used a cluster bomb design in which 106 four-pound submunitions were used to obtain a more efficient spread of the bacterial aerosol. No further trials were carried out in Scotland, but development of the anthrax bomb continued during the Second World War in co-operation with the United States and Canada. In May 1943 a compendium of all the British data was passed over to both the United States and Canada. Gruinard Island was eventually decontaminated in 1987 and returned to legatees of the previous civil owner in 1990.

Canada was likely to become involved in biological warfare activities because of its connections with the British government and military organizations. Before the Second World War, Sir Frederick Banting, winner of the 1923 Nobel Prize for the discovery of insulin, lobbied strongly for Canada to take biological warfare seriously. Banting was killed in an air crash on 21 February 1941 en route to England, but by then he had helped to establish a Canadian biological warfare organization. In November 1941 a secret biological warfare committee called M-1000 was set up to oversee its activities. In January 1942 a joint meeting of British, Canadian and US scientists with "an imposing agenda" took place in Ottawa. As we shall see in later chapters, the links between these three countries in biological warfare activities were to far outlast the Second World War.

On the defensive side, the Canadians were particularly concerned about the possible use of rinderpest virus against their cattle industry, and in a successful co-operative venture with the United States on Grosse Isle, an isolated island just thirty-five miles from Quebec, managed to develop a vaccine during the war. Also on Grosse Isle, in spite of many difficulties, they managed to produce the quantities of virulent anthrax required for continued testing at the Canadian Suffield test site. Considerable efforts were also made at Suffield to test weapons based on botulinum toxin (code-named X). In one new type of weapon tested, "30,000 X-tipped cluster projectiles, exploded from a 500 lb bomb ... Declustering ... at about 3000ft. ... In every case about 10,000 square yards were contaminated to a density of from 1 to 5 darts per square yard." The darts were said to penetrate six inches deep into sheep through the service clothing in which they had been clad. Botulinum toxin is the most deadly toxin known and this system would have delivered the toxin straight into the victim's tissues.

Intelligence reports in the spring of 1944 suggested that Germany had added botulinum toxin to its weapons arsenal and that this weapon might be used against Allied troops when Europe was invaded. The British and Americans decided that this was unlikely, but the Canadians were far less sure and lobbied hard for the troops to be inoculated against the toxin. The Canadians had managed to produce quantities of a toxoid against botulinum toxin type A, but eventually it was agreed not to inoculate the invading troops. Part of the Canadians' concern, of course, was that if they had successfully tested a weapon for which they had an antidote, perhaps the enemy had too. If the enemy had the same offensive/defensive combination for a botulinum toxin, why would it not be used if the Germans were desperate to stop the assault on the beaches? The close connection between offensive and defensive biological warfare programmes was clearly illustrated and this pointed up the great difficulty of understanding the meaning of whatever intelligence might become available from foreign secret programmes.

In the interwar years the US military was sceptical about the possibility of biological warfare being waged successfully. However, in 1940–1 there were reports of Japanese biological attacks in China and of Germany experimenting with botulinum toxin, and attitudes began to change. An influential report in February 1942 considered biological warfare "distinctly feasible". Gradually, the United States established the basis for what was to become, after the war, a massive offensive biological warfare programme. By the end of the war it had developed an impressive anti-plant biological weapons capability using both fungal agents that could attack rice and synthetic chemicals that mimicked natural plant growth regulators. One of these, LN-8 (2,4-dichlorophenoxyacetic acid, 2,4-D) in its ester form, Agent Orange, was to be used extensively in American biological warfare in Vietnam during the cold war.

By the time World War II ended in August 1945, the Special Projects Division – the powerhouse of the biological warfare effort in the United States – employed "396 Army officers, 2466 Army enlisted men, 124 Naval officers, 844 Naval enlisted men and 206 civilians". Anthrax and botulinum toxin took precedence (along, later, with the anti-plant agents) over other possibilities, but many other anti-personnel, anti-animal and anti-plant agents were investigated. Great effort also went into trying to perfect means of mass-producing lethal agents:

> Pilot Plant No. 1, completed in October 1943, was designed for the production of botulinum toxin; Pilot Plant No. 2, completed in March 1944, for the production of anthrax spores. ... In Pilot Plant T-62, part of No. 2, agent material was separated, dried and prepared for munition filling.

Furthermore:

> Two more pilot plants were completed in 1945: No. 3, designated for the production of plant pathogens; and No. 4, for the production of the agents causing brucellosis and psittacosis.

By the end of the war, capabilities in the United States for anti-personnel biological warfare were fast nearing completion. The Vigo production plant, which was not yet operational when the war ended, was designed to manufacture and load 500,000 four-pound anthrax bombs per month.

The United States' offensive programme had acquired two lethal anthrax strains, M-36 Vollum from the UK and N-99 from Cornell University. The V1 and V2 strains later grown in the US at Fort Detrick were twice as virulent as the original M-36. The eventually preferred strain No. 55 was three times as virulent as V1. This strain was shown to be stable in bulk storage. Five types of botulinum toxin, A, B, C, D and E, were available. Only A and B were studied and A was chosen as the more powerful and less familiar in Europe and Asia. The favoured strain of *Brucella suis* (the cause of brucellosis) was a highly virulent type termed S1. It was shown that animals were infected via the respiratory route and that the agent remained stable when discharged in a cloud.

At the time the war in Europe ended, there was considerable discussion in the United States about attacking Japanese rice production using plant growth regulators and defoliants. If the war had continued into 1946, it seems likely that the United States would have had a lethal anti-personnel capability as well.

## conclusions

We can see that in less than half a century biological warfare had moved from a scientific possibility at the time when the causes of infectious diseases were discovered to a practical possibility through the considerable efforts made by a number of major states. Germany had attempted a crude anti-animal sabotage campaign in the First World War and Japan had carried out unsophisticated large-scale field tests in China in the 1930s and early 1940s. France had taken a more scientific approach and was testing anti-personnel weapons just before the German invasion in 1940. Taking an equally scientific approach, the British first developed an anthrax anti-animal weapon and then worked out how to most effectively attack people with an anthrax aerosol weapon. This weapon was further developed in co-operation with the United States and Canada and would have been in production by 1946 had the war not ended. The United States also developed effective anti-crop agents during the Second World War and could have used these against Japan's staple food crop, rice, if the war had continued.

In short, by the end of the Second World War it was clear that biological agents could be weaponized to attack humans, animals and plants – even if at that stage the weapon systems were not very efficient – and that a great deal more development was possible. A further conclusion to be drawn from the history of these offensive biological programmes is that gaining accurate intelligence about secret biological programmes was, and remains, very difficult. Information was sparse and often consisted of dubious rumours, and it was very difficult to differentiate between measures undertaken for defensive purposes (e.g. development of botulinal toxoid antidotes) and the development of offensive capabilities (e.g. production of botulinal toxin agents). The classic case was the entirely understandable misreading of German activities by other states. In the circumstances, it was not surprising that other states resorted to worst-case analyses and imagined that German capabilities mirrored their own. As we shall see in the next chapter, such difficulties and responses also characterized the half-century that followed the Second World War.

## references

1. Eitzen, E.M. and Takafuji, E.T. (1997) Historical overview of biological warfare. In F.R. Sidell, E.T. Takafuji and D.R. Franz (eds), *Medical Aspects of Chemical and Biological Warfare*, pp. 415–23. Washington: Office of the Surgeon General.

2. Unless stated otherwise, the history of biological warfare recounted in this chapter is based upon the detailed studies in Geissler, E. and van Courtland Moon, J.E. (eds) (1999) *Biological and Toxin Weapons: Research, Development and Use from the Middle Ages to 1945*. Oxford: Oxford University Press.
3. Prescott, L.M., Harley, J.P. and Klein, D.A. (1990) *Microbiology*. Dubuque, IA: W.C. Brown.
4. Pelczar, M.J., Chan, E.C.S. and Krieg, N.R. (1993) *Microbiology: Concepts and Applications*. New York: McGraw-Hill.
5. Dando, M.R. (1994) *Biological Warfare in the 21st Century: Biotechnology and the Proliferation of Biological Weapons*. London: Brassey's.
6. Faludi, G. (1998) Challenges of BW control and defense during arms reduction. In E. Geissler et al. (eds), *Conversion of Former BTW Facilities*, pp. 67–72. Dordrecht: Kluwer Academic.
7. Whitby, S.M. (2001) *Biological Warfare against Crops*. Basingstoke, England: Palgrave.

chapter three

# biological warfare 1945–72

## introduction

All the major victors of World War II – the United States, the Soviet Union, the United Kingdom and Canada – ended the war with vigorous offensive biological warfare programmes in place. Soon after the war ended, France re-established its offensive programme, which had been interrupted by the German invasion of 1940. In this chapter the evolution of the American, British, Canadian and French programmes will be followed, up to the agreement of the Biological and Toxin Weapons Convention in the early 1970s. The Soviet programme (which was to reach its zenith later in the century) and the offensive programmes of Iraq and South Africa will be discussed in Chapter 4. There have been numerous accusations that other countries had such offensive programmes in the second half of the twentieth century, but the seven mentioned above are the only ones about which there is definite information.

It should be understood that, though constrained to a no-first-use policy by the 1925 Geneva Protocol, until the agreement of the Biological and Toxin Weapons Convention in 1972 it was not illegal for these states to *develop* biological weapons. After 1972, the offensive programmes of the Soviet Union (a co-depositary state for the BTWC), South Africa (a state party to the convention) and Iraq (a signatory of the convention) were illegal under international law.

## the united states

At the Third Five-Year Review Conference of the BTWC in 1991, following the first Gulf war with Iraq, states parties considered what might be done to improve confidence that states were living up to their obligations under the convention. Among the measures agreed was a series of additions to the confidence-building measures (CBMs) agreed at the previous review in 1986. One of the additions was CBM Form F, which asked, among other things, for a declaration of past activities in offensive biological research and development. The United States took the opportunity to provide a detailed account of its activities from 1941 through to the official closure of the offensive programme by President Nixon in 1969 (which led on to the eventual agreement of the BTWC in 1972). We can therefore use this declaration[1] as the basis for a description of the American programme. In 1977 the United States also produced a detailed, two-volume account, *US Army Activities in the US Biological Warfare Programs*,[2] which we can use to supplement the CBM declaration.

The US CBM declaration divides its history of the offensive programme into a series of time periods: 1941–6 (during the Second World War); 1946–9; 1950–3; 1954–8; 1959–62; 1963–8; and 1969–72. Volume 1 of the history of the US Army's activities helpfully provides summary title headings for its account of the postwar periods:

| | |
|---|---|
| 1946–9 | Research and planning years after World War II |
| 1950–3 | Expansion of the BW programme during the Korean War |
| 1954–8 | Cold war years – reorganization of weapons and defence programmes |
| 1959–62 | The limited war period – expanded research, development, testing and operational readiness |
| 1963–8 | Adaptation of the BW programme to counter insurgencies – the Vietnam War years |
| 1969–72 | Disarmament and phase-down. |

The second volume is a collection of detailed appendices to the first. For the brief overview of the postwar programme here we shall use the same time periods as the CBM.

As we saw in Chapter 2, the United States made significant progress in its offensive biological weapons programme during the Second World War. A final report on this work to the Secretary of War stated that the most important accomplishments included: "1. Development of methods and facilities for the mass production of micro-organisms and their products. ... 4. Production and isolation, for the first time, of a crystalline bacterial toxin. ... 10. Information on the effects of more than 1000 different chemical agents on living plants. 11. Studies of the production and control of certain diseases of plants." From this basis, the programme moved on to its postwar research and planning phase.

## 1946–9

During this period no production of agents was carried out on the scale required for operational readiness and no facilities were available for such work. The Vigo Ordnance Works production plant at Terre Haute, Indiana, which had been constructed during the war, produced nothing other than simulants before the end of the war and it was subsequently sold off to the Pfizer company for commercial use. The offensive biological weapons programme during this period concentrated on anti-personnel agents but also involved work on anti-animal and anti-plant agents. The agents recorded in the CBM return are shown below:

### *anti-personnel*

- *Brucella suis*
- *Pasteurella tularensis* (tularemia)
- Q-fever
- Venezuelan equine encephalitis (VEE)
- *Bacillus anthracis*
- *Staphylococcus* enterotoxin B

### *anti-animal*

- rinderpest
- hog cholera
- Newcastle disease virus

*anti-plant*

- wheat stem rust
- rye stem rust
- rice blast

The return noted that the small anti-animal programme was closely linked to the anti-personnel programme because the effects of the agents on animals and humans were similar. The anti-crop programme was separate because of the agricultural science disciplines it required.

At this time there was some small-scale outdoor testing of simulants, but the period is significant for the building of an enclosed one-million-litre testing sphere at Camp (later Fort) Detrick. This was the largest such testing sphere in the world and allowed the study of explosive munition tests with some of the pathogens under investigation. The "retaliation-only" policy for chemical and biological weapons was subject to a high-level civilian and military review in 1949, and the President approved a continuation of the policy in February 1950.

## 1950–3

Despite the reaffirmation of the retaliation-only policy, another substantial review in 1950 recommended that a biological weapons production facility, should be established, agent and munitions field tests should be carried out and all aspects of the research programme should be expanded. The programme was obviously to be enhanced in this Korean War period.

The US Army Chemical Corps assumed prime responsibility for carrying out these recommendations. The Army was authorized to build a production facility, at Pine Bluff Arsenal, Arkansas, to manufacture *Brucella suis* (brucellosis or undulant fever) and *Pasteurella tularensis* (tularemia). Anti-crop production sites were selected at Rocky Mountain Arsenal and Beale Air Force Base. A limited biological weapons retaliatory capability was said to have been achieved by the Air Force in 1951, when an anti-crop bomb was put into production. The first open-air sea tests were carried out using simulants against US naval ships off Norfolk, Virginia, and simulants were also released off San Francisco to discover how large a land area would be affected.

Yet in spite of the expansion of the programme, there were high-level worries about the quality of its management. In June 1953 the Secretary of Defense expressed concern over the state of CBR (chemical, biological, radiological) readiness. The Pine Bluff biological weapons anti-personnel agent plant, for example, was scheduled to go into production in October 1952, but production did not start until December 1953 and readiness to meet estimated production requirements was only achieved in the spring of 1954. The total cost of the plant was US $90 million.

## 1954–8

The Pine Bluff Arsenal was only able to produce *Brucella suis* (the causal agent of undulant fever) in the spring of 1954. Large-scale production of the lethal agent *Pasteurella tularensis* (as the agent causing tularemia was then called) began a year later. Nevertheless, major investments in the programme continued, Fort Detrick being expanded by US $15.6 million's worth of laboratory construction between August 1954 and July 1958.

The critical change during this period, however, was in high-level policy.[3] Reacting in part to Soviet statements in early 1956 to the effect that chemical and biological weapons would be used in future wars for purposes of mass destruction, US official policy was changed: "a revised BW/CW policy was formulated to the effect that the U.S. would be prepared to use BW or CW in a general war to enhance military effectiveness. The decision to use BW or CW was reserved for the President." In short, the former retaliation-only policy, based on America's signature of the 1925 Geneva Protocol (not ratified until 1975, but still, it could be argued, binding on them as a signatory), was abandoned.

A major symposium convened by the Defense Science Board at the headquarters of the Rand Corporation examined the political and military impact of biological weapons and resulted in a set of recommendations to specify weapon system requirements, increase the research effort, develop doctrines for use of biological weapons and even attempt to gain public support for biological and chemical weapons. Eventually, a new Ad Hoc Committee on Biological and Chemical Warfare was established to prepare an action programme based on these recommendations. Contrary to a commonly held view that biological weapons have never been seen to have great military value, the US military at this stage was clearly bent on producing effective biological weapons for use in major warfare without the constraints of a retaliation-only policy.

## 1959–62

In mid-1959 the Secretary of Defense was told that the available biological weapons were out of date and needed modernization. By late October of that year the Chief Chemical Officer was instructed to prepare an expanded five-year programme. The anti-crop programme, in which the Air Force had lost interest during 1957 and which was phased out, was revived and by the end of 1959 the Chemical Corps mission was described as having "reached a height of emphasis unprecedented since WWII".

In this period, when the Soviet Union was resorting to limited harassment tactics, for example in Berlin, and military thinking was turning towards the possibility of small-scale conflicts without nuclear exchanges, there was a perceived need for means of controlling conflicts and keeping casualties to a minimum. "Controlled temporary incapacitation, therefore, became a ... weapons objective, and CW and BW weapons offered the most promising technical possibilities." The US Army report thus notes that the biological weapons programme was switched to emphasize incapacitation.

When the Kennedy administration came to power a thorough reassessment of biological and chemical weapons under a project numbered 112 was undertaken. When the study was finished, the Joint Chiefs of Staff estimated that obtaining a complete spectrum of chemical and biological weapons capabilities would cost about US $4 billion. A Project 112 Working Group prepared a comprehensive plan of tasks, actions and target dates. Within the plan US $20.1 million was approved to expand and modify the production plant at Pine Bluff. Development work on Q-fever and tularemia allowed their standardization as biological weapons agents. Also, the testing programme was enhanced by activation of a chemical and biological testing organization at the Deseret Test Center, Fort Douglas, Salt Lake City.

## 1963–8

During the Vietnam War period the biological weapons programme was guided by the requirements set out in Project 112. The expansion of the production facilities at Pine Bluff was begun in 1964 and completed in 1967. The plant produced several different agents. Various types of biological weapons munitions were delivered to Pine Bluff and were filled and stored there. Production of anti-crop agents was accelerated in 1963 and continued through to 1968.

The United States used chemical herbicides extensively in Vietnam both to defoliate jungle terrain and to destroy crops. The Army account states that this did not constitute biological warfare. However, the material used was a synthetic analogue of a plant hormone. It would be categorized as a mid-spectrum agent today and would therefore fall under the prohibition of the BTWC.

From 1964 onwards considerable work was carried out on *Staphylococcus* enterotoxins in order to be able to cause severe short-term incapacitation. Because much smaller amounts of these enterotoxins were needed to cause incapacitation than of standard chemical agents, they looked attractive to the weaponeers who eventually believed weaponization was possible.

## 1969–72

On 25 November 1969 President Nixon announced a major policy change on chemical and biological weapons. He renounced the first use of lethal and non-lethal incapacitating chemicals and stated that he would submit the 1925 Geneva Protocol for ratification by the Senate. Regarding biological weapons, he renounced the use of "lethal bacteriological (biological) agents and weapons and all other methods of biological warfare". On 14 February 1970 he followed this up by renouncing use of biological toxins, whatever their means of production. The Army account notes that, though there had been several major reviews of US policy on chemical and biological weapons, "the origin of the policy change dates from criticism of U.S. application of chemical herbicides and riot control agents in the Vietnam war beginning in the mid-1960s". So, adverse public opinion against these weapons was important in bringing the massive programme to an end.

The 1997 Chemical Weapons Convention specified that the huge accumulated stocks of chemical weapons, including the very dangerous nerve agents, must be destroyed. Because it was technically very difficult to destroy the lethal nerve agents safely, some people assume that it is also difficult to destroy biological agents, but this is not true. Destruction of the stocks of anti-personnel agents and munitions was carried out between 10 May 1971 and 1 May 1972 at

Pine Bluff Arsenal. The biological agents were essentially disposed of by heating them over a three-hour period. Destruction of anti-crop agents was carried out at the three storage sites (Rocky Mountain Arsenal, Beale Air Force Base and Fort Detrick) between 19 April 1971 and 15 February 1973. The agents were destroyed by chemical treatment and heating.

## assessment of the US offensive programme

Volume 2 of the US Army's account summarizes the results of the offensive programme activities (in "Annex C: Biological Warfare Research and Development"). For *anti-personnel* agents, it indicates that a wide range of highly infectious micro-organisms and extremely poisonous toxins were studied, research efforts being directed towards "selection and preservation of the most virulent strains, establishing human dosages, enhancing storageability, and survival when released as an aerosol". Technology for the large-scale production of the most promising agents was also developed and simulant organisms were selected and used in field testing. Nevertheless, the twenty-six years of the offensive programme resulted in the standardization of only eight anti-personnel agents (see, for example, Figure 3.1). Another annex (D) reports that the following agents were produced at Pine Bluff between 1954 and 1967: *Brucella suis*; *Pasteurella tularensis*; Q-fever; VEE; *Bacillus anthracis*; botulinum toxin; and staphylococcal enterotoxin. *Anti-crop* agent research included strain selection, evaluation of nutritional requirements, development of optimal growing conditions and harvesting techniques along with preparation in a form suitable for dissemination. There was also extensive field testing to assess the effects of the agents on crops. In the end, of the many anti-crop agents screened, five were standardized. These included wheat and rye stem rust and rice blast. *Anti-animal* research began during World War II and was initially directed towards discovering means of protecting livestock. Vaccines were developed against rinderpest (a deadly cattle disease) and Newcastle disease, which affects poultry. Two field tests were carried out using hog cholera virus and Newcastle disease, but the whole programme was terminated in 1954.

Early research and development on munitions in the United States started with adaptations of the burster-type bombs developed by Britain during the war. However, the work was

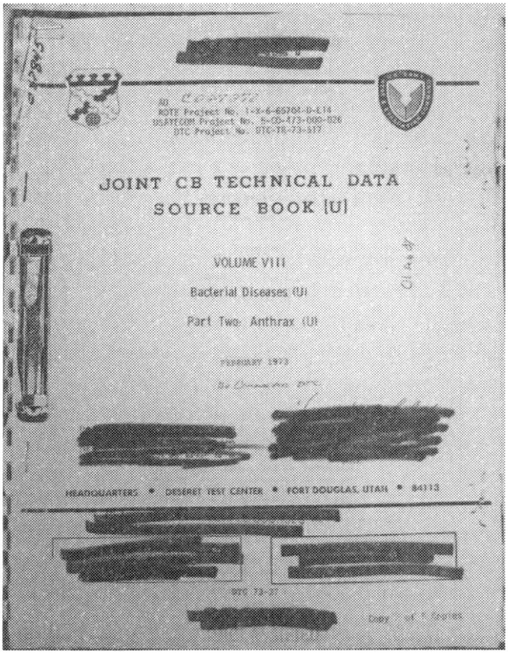
[redacted]

JOINT CB TECHNICAL DATA
SOURCE BOOK (U)

VOLUME VIII

Bacterial Diseases (U)

Part Two: Anthrax (U)

[redacted]

HEADQUARTERS • DESERET TEST CENTER • FORT DOUGLAS, UTAH • 84113

[redacted]

[redacted]

**Figure 3.1** *Front cover of the data collation on the US anthrax agent. One of a series of Joint CB Technical Data Source Books on biological agents produced in the US and later declassified.*

extended "to improve burster type munitions, submunitions, gas expulsion bombs, various types of line spray tanks and highly specialised projectiles and generators as well as insect vectors". The research involved, for example, optimizing munition configurations, testing performance and developing hardware production and filling technology.

Little wonder then that many of those involved were reportedly not at all happy with President Nixon's decision to abandon what appeared to be, despite the technical difficulties and setbacks, a successful outcome to a quarter of a century's work. The story appears to have been put about that the programme was abandoned primarily because the weapons were unreliable – for example, being much dependent on the weather. However, an equally cogent reason could have been the increasing realization that very few nations could acquire nuclear weapons of mass destruction (because of the difficulty of producing the fissile material) whereas many states could develop biological weapons of mass destruction. Why not then renounce biological weapons, spread the idea that they were of little use, and even negotiate a total ban on them?

One point that needs to be clearly understood is that the utility of biological agents was rigorously tested in the US offensive programme. Annex E in Volume 2 of the Army's account is concerned with the details of testing. It points out that testing is an integral part of research and development and is primarily concerned with obtaining empirical data in order to evaluate postulates made in the laboratory and in mathematical models. The categories of testing considered were laboratory (small-scale), closed chambers (medium-scale) and open-air field tests (large-scale). For the purpose of confirming that biological weapons could cause casualties on the scale of a weapon of mass destruction, large-scale (open air) field tests were obviously required. Tests were carried out in the anti-personnel programme both with "simulant" microbes considered non-pathogenic to humans and with pathogens, and also with anti-crop agents.

Annex E has a series of tables (see Table 3.1 here), running to sixteen pages, listing the biological field testing known to have been carried out in the United States and elsewhere during the offensive programme. A further annex (K) deals with testing of agents on

**Table 3.1** Titles of tables listing US biological field testing*

1. Anti-personnel biological simulants involving the public domain
2. Anti-personnel biological simulants not involving the public domain
3. Non-biological simulants/air diffusion involving the public domain
4. Anti-animal non-biological simulants involving the public domain
5. Anti-personnel pathogenic agents
6. (Unsubstantiated) anti-personnel pathogens not involving the public domain
7. Anti-crop pathogenic agent involving the public domain
8. (Unsubstantiated) anti-crop biological agents involving the public domain
9. Anti-crop pathogenic agent not involving the public domain

* From Department of the Army, op. cit.

human volunteers, particularly on members of the Seventh Day Adventist Church in Project Whitecoat. These church members refused to serve in the Army as combatants when drafted but were willing to serve as test subjects in the programme. The annex has fourteen pages listing the tests carried out on hundreds of such volunteers. It can be seen, therefore, why the people involved in the programme had a solid empirical basis for their views on the potential effectiveness of biological weapons.

## the united kingdom

As might be expected from the state that originated the BTWC and is one of the three co-depositary states with special responsibility for the convention, the United Kingdom, like the United States, attempted on its CBM Form F return to give an account – from the beginning – of its offensive programme.[4] The return explains that the UK had a programme intended to provide a capability for retaliation in kind should the UK ever be attacked with biological weapons and that the programme, which started in the 1940s, ceased in the late 1950s.

The return then briefly recounts the history of British concerns about biological warfare from the 1920s onwards and deals with wartime activities (of about forty-five people at most) in the biology department at Porton Down and the links between this group and the US and Canadian programmes as the war progressed. The account confirms that the stockpile of five million anthrax-laced cattle cakes developed to provide a retaliatory capability during the war was largely destroyed immediately after the war by autoclaving and burning. A few boxes of cakes kept at Porton Down in the culture collection were finally destroyed in 1972 when the BTWC was signed.

The account states that, though some research on offensive aspects continued after the Second World War, work on an offensive capability had been abandoned by 1957. Under the following section of the return, which deals with past defensive biological research and development programmes, the offensive nature of some of the work carried out is clearly indicated:

> in the period 1948–1955 field trials with pathogens were performed on the high seas off the Bahamas and off the Scottish coast, initially

> to determine the feasibility of conducting trials at sea and latterly to acquire data on the behaviour of aerosols under realistic conditions.

The return goes on to state that "such work was begun during the period when offensively motivated R &D was ... being pursued".

Though the UK generally operates a thirty-year delay rule for the release of official documents and may retain some papers for longer than that, a great deal of information about the period of the offensive programme has been released into the public record at the National Archives at Kew in London. This has allowed scholars to investigate what happened, and why it happened, in much more detail.[5] Three aspects of this history are of particular interest. First is the gradual evolution over the period of the programme from a concentration on standard munitions such as cluster bombs for the delivery of the agent to the realization that huge areas could be put under threat by delivery of the agent as an aerosolized spray from a moving vehicle such as a plane. Secondly, there was intense tripartite co-operation between the US, the UK and Canada in the offensive programme. Third and most significant, however, is the evidence that biological weapons were viewed as on a par with nuclear weapons at the end of the Second World War. Only when the UK obtained its own nuclear systems did interest in biological weapons decline. With the constraints of a limited defence budget, it is understandable that the UK's offensive programme tailed off faster than that of the United States. The same truth is quite evident in the history of the French postwar offensive biological weapons programme.

A final point worthy of emphasis, in view of current worries about military interest in developing so-called non-lethal chemical weapons such as the fentanyl-like agent used to break the 2002 Moscow theatre siege, is that the UK made a vigorous effort to find a non-lethal, incapacitating chemical agent. Such an agent would have acted to disrupt the chemical transmitter messengers of the brain and would therefore now be classed as a mid-spectrum agent subject to the prohibition embodied in the BTWC. The scientists at Porton Down were already investigating the subject because of ongoing US work, but the Cabinet actually made the decision in 1961 to restart offensive work with the aim of finding such an agent. At this time the United States weaponized BZ (3-quinuclidinyl benzilate), but it was never used and in any case was regarded by the British as too unpredictable. Despite much scientific work through the 1950s and 1960s, there was at that time insufficient understanding of the nervous system to give the quest for an effective, non-lethal, incapacitating chemical any real chance of success.

## canada

Canada's declaration under Confidence-Building Measure F, regarding past offensive programmes, states that Canada did not engage in large-scale production, stockpiling or weaponization of biological weapons agents and that there was no work aimed at determining the suitability of specific agents as weapons.[6] However, this statement comes at the end of a paragraph that begins by listing some of the work that Canada *was* involved in:

- studies of improved procedures for production of certain toxins (e.g. botulinum and diphtheria);
- studies on the use of insects as vectors for pathogenic bacteria;
- test and evaluation of munitions, including performance in cold weather;
- studies of weapon-produced aerosols of potential BW agents;
- fundamental work related to field trials, dealing with the dispersion and properties of solid particulates, preparation of finely divided solids for munitions charging and sampling of toxic particulates;
- development of tissue culture processes for large-scale cultivation of viruses;
- development of *Mallei mallei* and *Mallei pseudomallei* as new potential BW agents.

This is very interesting when trying to understand just what was going on at the time Canada was co-operating with the US and UK on offensive biological research and development. The period of the work is stated in the declaration to have been between 1946 and 1956. As in the UK, a good deal of the official documentation produced during the period of the offensive programme has been put into the public record and since subjected to scholarly analysis.[7] It can be seen from the list above that, although Canada may never have possessed biological weapons, it certainly did a great deal to assist the British and American offensive programmes. Canada, in fact, played a full part in the tripartite system and its annual meetings and specialist working groups.

Canada had been co-operating with the British during the Second World War before the entry of the United States into the conflict. During the war Canadian scientists had produced a botulinum toxoid (A & B) which might well have been used to immunize troops against the toxin had it been judged that the Germans were likely to employ it during the D-Day landings in Normandy. In the event, the toxoid was not needed because Allied leaders concluded that the toxin would not be used. A joint US–Canadian project at Grosse Isle, Quebec, had also resulted in the production of an effective vaccine against the dangerous animal agent of rinderpest disease.

Though Canada may have been a junior partner in the postwar tripartite system, it had one very great advantage in its Suffield test site, where large-scale chemical and biological field tests could be carried out. Such tests were done between 1945 and 1950 on both a four-pound bomb and a five-hundred-pound cluster bomb. Suffield had a seven-hundred-square-mile operational area and a one-mile-square area with a rodent-proof fence considered highly suitable for safe biological weapons agent testing. The Grosse Isle site was also reactivated at the end of the 1940s and used in co-operation with the United States in testing a large number of potential anti-animal agents during the early 1950s. Canadian involvement in US field testing appears to have continued through the 1960s until the end of the US programme.

By the end of the 1960s, Canadian diplomats were much involved in moves within the United Nations system to find means of agreeing a biological weapons convention. Despite this, the Canadian government was taken by surprise by President Nixon's sweeping renunciation of biological weapons. However, within a few weeks Canada moved to support the American position and has been a strong supporter of the BTWC since its agreement in 1972.

## france

In France the offensive biological weapons programme that had been interrupted by the 1940 German invasion was quickly re-established after the war. Only very recently have scholars had access to data about the programme.[8] The programme flourished between 1947 and 1956, but from 1956 through to 1972 there was a budgetary redistribution towards nuclear weapon systems and the biological weapons programme went into decline. Throughout the period of the offensive programme, however, France, as the depositary state for the 1925 Geneva Protocol, had a clear retaliation-only policy for its biological weapons.

Though dwarfed by the US programme and smaller than the postwar British programme, the French programme built carefully on the considerable advances made before the Second World War and gave France a great deal of insight into the potential dangers. It is significant that France objected strongly to the lack of proper verification measures in the BTWC and did not immediately become a state party to the convention.

## conclusions

The conclusions we can draw from this immediate postwar history might be summarized under three headings: resources; results; restraints. It is quite clear that the victorious countries came out of the Second World War with intact offensive biological weapons programmes and considered them so important that they invested major resources in their continuation. In the postwar austerity period in the UK, for example, a new purpose-built laboratory was constructed at Porton Down costing several million pounds. The building was thought to have been one of the largest brick-built structures at that time.

In terms of results, the crude initial work carried out during the war was supplemented by years of careful scientific investigation and large-scale field trials which demonstrated that diseases could be deliberately and effectively used. Short of use in all-out warfare, what has been called "medicine in reverse" was worked out and tested to the weaponeers' satisfaction. Yet there were restraints. The conceptual understanding of viruses only really firmed up in the 1950s, so most studies were on bacterial and fungal agents. Of immense significance for the future, Watson and Crick demonstrated the double helix structure of DNA – the hereditary material – in the early 1950s. It was to take until the early 1970s for genetic engineering technology – the movement of functional genes between different species – to be shown to work. So at the time the BTWC was being agreed the foundation for the current revolution in the life sciences, with all that portends for good and ill, was also being put in place. The stage was thus set for the application of this new knowledge in the illegal Soviet programme of the later cold war years.

## references


1. United States (1992) *Confidence Building Measures.* In DDA/4-92/ BWIII, pp. 252–428. United Nations, 30 April.
2. Department of the Army (1977) *U. S. Army Activity in the U. S. Biological Warfare Programs.* Volumes 1 and 2. Washington, 24 February.
3. See van Courtland Moon, J.E. (in press) The US BW program: dilemmas of policy and preparedness. In M. Wheelis, L. Rozsa and M.R. Dando (eds), *Deadly Cultures: Bioweapons from 1945 to the Present.* Cambridge, MA: Harvard University Press.
4. United Kingdom (1992) *Confidence Building Measures.* In DDA/4-92/ BWIII, pp. 213–52. United Nations, 30 April.
5. See, in particular, Balmer, B. (2001) *Britain and Biological Warfare: Expert Advice and Science Policy.* Basingstoke, England: Palgrave.
6. Canada (1992) *Confidence Building Measures.* In DDA/4-92/ BWIII, pp. 30–82. United Nations, 30 April.
7. See Avery, D. (in press) Cosmic top secret: Canada, Alliance co-operation and the biological arms race. In M. Wheelis, L. Rozsa and M.R. Dando (eds), *Deadly Cultures: Bioweapons from 1945 to the Present.* Cambridge, MA: Harvard University Press.
8. See Lepick, O. (in press) The French biological weapons program, 1947–1972. In M. Wheelis, L. Rozsa and M.R. Dando (eds), *Deadly Cultures: Bioweapons from 1945 to the Present.* Cambridge, MA: Harvard University Press.

chapter four

# biological warfare 1972–2004

## introduction

During the 1990s, as concerns about biological weapons resurfaced in the public domain following the first Gulf war of 1991, there was considerable speculation about which countries might be harbouring offensive biological weapons programmes. In its publication, *Proliferation of Weapons of Mass Destruction: Assessing the Risks*, the US Congress Office of Technology Assessment in 1993 gave a list of countries and then a list of six different publications that indicated whether the author thought particular countries had such a programme.[1] As can be seen from Table 4.1, there was considerable agreement between the various publications about the presence of offensive programmes in some states. However, it has to be understood that the evidence for these assertions has not been rigorously and independently analysed and, following the controversies over the status of the Iraqi programme in the late 1990s and early 2000s, we are all very aware of the difficulties even the intelligence services of the major powers have in making valid estimates of the status of foreign programmes.

What we do have is definite evidence of three state-level offensive biological weapons programmes after the agreement of the BTWC in 1972. The Iraqis had a biological weapons programme that, despite their efforts to hide it by lies and dissimulation, was pinned down by the United Nations Special Commission (UNSCOM) inspectors by 1995. South Africa also had an offensive biological

**Table 4.1** Biological weapons programmes suspected in various publications*

| Country | Number of citations | Percentage |
|---|---|---|
| Libya | 5 | 83 |
| North Korea | 5 | 83 |
| Iraq | 5 | 83 |
| Taiwan | 4 | 67 |
| Syria | 4 | 67 |
| Soviet Union | 4 | 67 |
| Israel | 4 | 67 |
| Iran | 4 | 67 |
| China | 4 | 67 |

* From OTA, op. cit.

weapons programme in the later stages of the white apartheid regime, but the programme was terminated before the regime change in the 1990s. The Iraqi programme might best be compared to the kind of programme available to the developed countries in the years following the Second World War. The South African programme was on a smaller scale and seems to have been devoted to finding means of assassinating the apartheid regime's enemies. The Soviet programme in the later stages of the cold war was of a quite different order. A vast programme, which has yet to be fully analysed because of the non-release of official documentation, set about using the increasing capabilities of the life sciences for the purpose of creating an effective military programme.

From an analysis of these three different programmes we can, perhaps, gain an overview of what might be going on in other countries for which there are good grounds for suspicion and get a picture of the possible current status of biological weapons development at the turn of the new century. The Soviet programme also provides us with a means to begin considering what might happen if we are unable to prevent states undertaking major offensive biological weapons programmes in the coming decades.

## iraq

Iraq acceded to the 1925 Geneva Protocol in 1931 with a reservation that gave it the freedom to use chemical or biological weapons in retaliation.[2] Iraq also signed the BTWC in 1972 (which bound it to respect the convention) but did not ratify its signature until it was required to do so by UN Security Council Resolution 687 (1991) after it was defeated following its invasion of Kuwait. Iraq's offensive biological weapons programme was therefore illegal and followed its flagrant breach of the 1925 Geneva Protocol by attacking Iran first with chemical weapons during their long war in the 1980s.

The state of Iraq was created in 1921, following the First World War, and was administered by the UK under a League of Nations mandate until the country became independent in 1932. Iraq was one of the founder members of the United Nations in 1945 and was a monarchy until 1958, when a republic was established. The Arab Ba'ath Socialist Party seized power in 1968. With Saddam Hussein as head of government, the quest for weapons of mass destruction soon began, small quantities of chemical agents being produced by 1971. Iraq has stated that its search for biological capabilities began shortly thereafter, in 1974, but that this initial work produced nothing of significance. It is not certain whether this was so because little reliance can be placed on Iraqi testimony without corroborative evidence.

According to the Iraqis, a biological programme began again in earnest in 1985, during the war with Iran. The Iraqis say that two agents – anthrax and botulinum toxin – were selected as candidate agents because other countries had used them for biological weapons purposes and they were relatively easy to produce. Initial work was said to be limited to studies of pathogenicity, toxicity and laboratory-scale production. In the mid-1980s there was also a separate stream of work investigating the use of wheat smut as an economic weapon against crops. After an expansion of the programme, further work from 1987 onwards involved bench-top fermentation experiments with anthrax and botulinum toxin and experiments on a wide range of animals, for example to study inhalation of the agents. Also at this time work was carried out on the bacterium *Clostridium perfringens* and on fungal toxins such as aflatoxin. Field trials of a crude dissemination device began in 1988 and were followed by more sophisticated biological bombs and spraying devices. It appears that at this stage, at the end of 1987, plans were made to purchase the apparatus needed for large-scale agent production. The main production site at Al Hakam – in the desert about fifty-five kilometres south and west of Baghdad – was rapidly and secretly constructed between 1988 and 1990. When foreign suppliers could not be found to provide fermenters, the equipment was transferred from a civilian facility in Iraq.

Iraq has stated that between January 1989 and August 1990 it produced 13,600 litres of concentrated botulinum toxin at Al Hakam and five thousand litres at another site. Anthrax production began at Al Hakam in June 1990 and by August 170 litres had been produced. Anthrax was also produced at another site. Agents such as aflatoxin were produced at different sites and work was also begun on other agents such as ricin toxin extracted from castor oil beans. Studies were also made of the best way to dry anthrax spores. Iraq tried to obtain drying equipment, but this effort was thwarted when an export licence could not be obtained by the supplying company.

As the prospect of war with UN forces increased, work on weaponization of biological agents accelerated and aerial bombs, missiles, rockets and sprayers were produced (Table 4.2). Iraq has also stated that authority to launch chemical and biological weapons was pre-delegated to regional commanders should Baghdad be hit

**Table 4.2** Some of Iraq's biological weapons*

| **Delivery system†** | **Agents** |
|---|---|
| Al-Hussein missile warheads | 16 filled with botulinum toxin |
| | 5 filled with anthrax |
| | 4 filled with aflatoxin |
| R-400 aerial bombs | 200 produced |
| | 100 filled with botulinum toxin |
| | 50 filled with anthrax |
| | 7 filled with aflatoxin |
| Aerosol generators/helicopter spray system | 12 devices produced |
| | Successful field testing with simulant |
| 155-mm artillery shell | 4 filled with ricin and used in a trial |
| Fragmentation weapon | Said to have been trialled with *Clostridium perfringens* |

* From Pearson, op. cit.

† Not all systems listed and there are questions on some of the numbers.

by nuclear weapons during the likely war. However, UNSCOM noted that this did not exclude alternative forms of use and did not constitute proof of a retaliation-only policy.

It is therefore certain that Iraq went into the 1991 Gulf war with some biological weapons available for use. Though the weapons were relatively crude and used liquid rather than dried and milled agent, it seems unlikely that the next step – of producing a more sophisticated and effective dried agent – was something that would have taken the Iraqis very long to perfect. What is also clear is that Iraq tried very hard to conceal its offensive programme for biological weapons from UNSCOM and it was not until 1995 that a reasonable picture of its activities began to be uncovered. The history of the later 1990s, and the early part of the new century until Iraq's defeat in the second Gulf war, remains shrouded in mystery and controversy. At present it does not appear that Iraq was engaged in producing and stockpiling agents, but its refusal to clarify what it was doing still leaves many believing that Saddam Hussein's regime intended to carry on with the programme if and when conditions permitted. It should also be understood that Iraq had a great deal of experience with chemical weapons and pursued not a Western-style approach of production and stockpiling but rather production and rapid use (which required a far less pure agent) and it is possible that a similar approach could have been taken towards biological weapons. Given such an approach, the task of international inspectors attempting to clarify the history and status of the programme would prove to be much more difficult, since Iraq would not have carried out many anticipated activities.

Controversies aside, what can we learn from the Iraqi biological weapons programme? The main lesson is that a medium-sized country without great scientific and technical resources was, within a few years, able to reach the stage of weaponizing a range of deadly biological agents including highly lethal anthrax and botulinum toxin. The possibility cannot be dismissed that some or all of the other countries listed in Table 4.1 have proceeded down the same path with equal success.

## south africa

An official account based on original documents is not yet available for the South African programme, but information can be found in material presented to the Truth and Reconciliation Commission, the trial of Wouter Basson, the head of the programme, and in the personal recollections of some of the scientists who participated.[3] The programme was highly secret and operated through front companies so there is some doubt whether any remaining documents that may be recovered will be of use to historians.

Most South Africans only learned about the programme through limited press reports when the front companies were privatized in 1992. Only in 1996, as allegations of fraud by those involved arose, did official investigations elicit further details about 'Project Coast'. Then in 1998 the Truth and Reconciliation Commission began to hear evidence from the scientists and military and a clearer picture began to emerge. This picture was of a small-scale programme carried out by an administration devoted to retaining white power at almost any cost. Those involved tried to develop and produce novel crowd control and assassination weapons for use against the regime's enemies, despite South Africa being a party to the BTWC.

South Africa's production facilities for chemical mustard gas were closed down at the end of the Second World War and it was not until 1981 that a secret chemical and biological programme was restarted. It would appear that the principal initial reason for the chemical programme was to provide protection for South African forces fighting in Angola should chemical weapons be used. A secondary reason was to find new forms of crowd control for the South African police. It has been suggested that the largely offensive biological weapons programme began as a means for testing novel chemical agents. The biological warfare facility at Roodeplaat Research Laboratories (RRL) functioned under the guise of a legitimate (front) company.

Some testing did take place at RRL, but the scientists involved have testified that the biological programme was intended to supply the military and police with covert assassination weapons. There is documentary evidence that colourless, odourless agents that were not traceable in post-mortems were sought. No great differentiation was made between chemical and biological agents – either would do. Although it was a small programme, the total costs of RRL including construction, running and selling-off were a not inconsiderable ninety-eight million rand. As a front company outside regular military organization, it offered the scientists working there better terms and conditions than if the operation had been strictly military. This seems to have been important in motivating many of the scientists who took part. A small amount of commercial work was also undertaken to maintain the fiction that RRL was a commercial company.

It appears that the idea for RRL and its activities came in the early 1980s from Wouter Basson's joint interest in, and discussions of, chemical and biological warfare with Daan Goosen, an academic. Goosen became the first managing director of the organization and has stated that as early as 1983 he supplied Basson with a live black mamba snake and a vial of its venom. He believed that this was used for an assassination and was clear that he and his colleagues' task was to supply biological assassination weapons resulting from their work at RRL.

Roodeplaat Research Laboratories was built on a seventy-hectare site just north of Pretoria in 1986. The final buildings and laboratories were completed in 1988, just a few years before the privatization of the company and the end of the offensive programme. So the period over which the work was carried out was very short. One of the scientists involved has stated that the kinds of things supplied were: beer contaminated with botulinum toxin; sugar contaminated with *Salmonella*; and chocolates laced with anthrax and botulinum toxin. Another scientist genetically modified the bacterium *Escherichia coli* to express the dangerous epsilon toxin of *Clostridium perfringens* (gas gangrene). Another project, ultimately unsuccessful, was to find an anti-fertility vaccine. Goosen believed that, had it succeeded, the vaccine would have been administered to black women without their knowledge.

Before the organization was closed down, there were certainly plans for a major expansion which included freeze-drying, storage and P4 laboratory facilities for dealing with highly virulent strains. A scientist involved has said that he believed the intention was to go into large-scale production of agents like anthrax, *Brucella* and *Salmonella* and also botulinum and tetanus toxins. None of this came to fruition, but it has to be asked what might have been done if the apartheid regime had not fallen. Much of the documentation of RRL was said to have been destroyed on privatization, but it appears that not all of the cultures were destroyed, for an attempt was made by Goosen to sell a collection of cultures and other products to the United States. Although this cloak-and-dagger operation eventually fell through, it leaves doubts about the thoroughness of the final closure of the programme.

The South African programme provides a very different example to that of Iraq but a functional and deadly one nevertheless. It also has to be said that since its installation in 1994 the new South African government has played a very positive role in attempts to strengthen the BTWC.

## the soviet union

It is widely agreed that in the latter part of the cold war the Soviet Union ran the largest offensive biological weapons programme of the twentieth century. The programme was expanded greatly following a decision of the Central Committee of the Soviet Communist Party in 1973 (the year after the agreement of the BTWC) and ran at least until 1992, when President Yeltsin acknowledged its existence. Questions remain today about the precise status of the programme because no official account of any credibility has been produced and some military establishments have remained closed to outsiders.

It appears that the programme was not completely hidden from the Western intelligence agencies during the cold war, but the scale of the programme was a considerable surprise when Vladimir Pasechnik defected to the West in 1989 and provided a great deal of reliable information. His defection was followed by that of Ken Alibek, who later produced his own account of what he knew of the programme. Alibek's book with Stephen Handleman was titled *Biohazard: The Chilling True Story of the Largest Covert Biological Weapons Program in the World – Told from the Inside by the Man Who Ran It*. It was published in 1999 and brought the programme much more to public notice. Other important participants who are less well known in the West have also published memoirs, and popular books such as Judith Miller and colleagues' *Germs: Biological Weapons and America's Secret War* and Tom Mangold and Jeff Goldberg's *Plague Wars: A True Story of Biological Warfare* have added to public knowledge. It has to be recognized, nevertheless, that a great deal of information about this programme is simply not available in the public domain. What we are therefore dealing with is a very partial account of an enormous, multifaceted endeavour. We do not even know, for instance, how many people were actually involved. At the programme's height it has been estimated that some 25,000–60,000 people were working in the programme, but it is not clear whether this estimate includes support staff or the military.[4]

Soviet interest in biological warfare is of long standing. As we saw in Chapter 2, early in the Soviet era Yakov Fishman, head of the Military Chemical Agency of the Red Army, prepared a report on the subject, and by the time of the Second World War the Soviet Union seems to have developed and tested a variety of BW systems. However, it is generally considered that the biological weapons programme during the war was of limited scope and that any production was on a small scale. The programme was expanded after the war but may well have been affected by the attack on orthodox biology during the ascendancy in official policy of the unfounded views of Lysenko. Certainly there was increasing concern about the backwardness of Soviet biology during the 1960s. By February 1966 the Soviet Council of Ministers had made the decision to strengthen the country's biological sciences and a further expansion occurred following the 1973 decision on the offensive programme.

The structure of the offensive programme was very complex, but it can be viewed more simply as comprising three parts: a military component, a political component and a civilian one. The military component was controlled by the Ministry of Defence's 15th Directorate and the political component reached right to the top of the Soviet system. What was unique was that the civilian component, called Biopreparat, appears to have been enormous, with at least twenty main locations. Although a supposedly civilian enterprise, Biopreparat was directed by the military and always had top-level military leadership.

The offensive programme had a major field-test site on Vozrozdenie Island in the Aral Sea. It had key research institutes such as those at Obolensk, Novosibirsk and Zagorsk (Table 4.3). Its production and storage facilities included those of Biopreparat at Berdsk, Omutninsk and Stepnogorsk, some of which carried out both research and production activities.

A major Biopreparat facility was the All-Union Institute of Highly Pure Biological Preparations set up in Leningrad in 1974. The institute eventually had some 3500 people working at three different sites and concentrated on weaponizing *Yersinia pestis*, the cause of plague, which is contagious from the first victim infected. By 1987 it was apparently capable of producing two hundred kilograms of weaponized material each week. The US programme did not weaponize contagious agents and this shows just one of the ways in which the Soviet programme went beyond what had been done before.

**Table 4.3** Some key research institutes in the Soviet programme*

| **Title** | **Place** |
|---|---|
| All-Union Scientific-Research Institute of Applied Microbiology | Obolensk, near Moscow |
| NPO "Vector" later called State Research Centre of Virology and Biotechnology (Vector) | Kol'stovo, Novosibirsk region |
| Institute of Experimental Hygiene | Kirov |
| Institute of Microbiology of the Ministry of Defence | Zagorsk |

* From Hart, op. cit.

Alibek suggests that research and development work was carried out on anthrax, brucellosis, tularemia, glanders and other agents that would have been expected from the US programme, but also on Ebola virus, Marburg virus, *Yersinia pestis* and smallpox.[5] Smallpox is a deadly contagious agent and Ebola, Marburg and the like were then recently discovered new and deadly viruses. This again shows that the Soviet offensive biological weapons programme had advanced to a new stage. Great emphasis was also placed on the ability to produce massive amounts of material so there was a very large standby production capability.

Tom Mangold recounts how Western inspectors visited the Berdsk facility under the Trilateral Agreement which it was hoped, in the early 1990s, would clear up suspicions about what the Russians were going to do under the new political regime. In one building the visitors saw four huge sixty-four-thousand-litre fermenters. In another building where construction had been stopped by Gorbachev in 1990, "the long, tall, central room had been prepared to hold four rows of 64,000 litre fermenters – with ten units in each row". If all forty had operated, these fermenters, the largest available in the Soviet Union, would have given a total capacity of 2,560,000 litres.

Though we do not have any official documentation on what the operational doctrine was in regard to biological weapons, with that level of preparation it has to be accepted that – at the very least – the Soviet Union had contingency plans for large-scale use of biological weapons in the unfortunate circumstance of a general war breaking out with the West.

So the programme in the Soviet Union had new and worrying features if we think it might foreshadow future offensive biological weapons programmes. Contagious agents were weaponized, new and deadly viruses were weaponized and production capacity was enormous. But these features are not the only cause for major concern. Although it is not easy to tell whether some research is for defensive or offensive purposes, as Judith Miller recounts in *Germs*, some of the open reports of the experiments carried out by Soviet scientists caused real concern in Washington.

We now know that the emerging capabilities being generated in the life sciences – genetic engineering – were being put to use in the Soviet offensive programme. The full details remain unclear, but it seems, for example, that attempts were made to render pathogenic bacteria like plague bacilli resistant to multiple antibiotics and to modify pathogens so that they would overproduce one of the body's natural chemical signalling molecules (bioregulators) and so disrupt its functions. Efforts were apparently made to produce chimera viruses by combining two different types into one. We shall revisit these issues in greater depth in Chapter 6, but clear already are the implications for the future. Should new offensive biological weapons programmes arise in years to come we should anticipate such use of burgeoning capabilities in the life sciences for malign purposes. The Soviet programme would then come to be seen as a harbinger of a terrible future posing enormous threats to human rights.

## allegations

It is difficult for intelligence agencies without good access to a foreign country to understand what is being done there and to know whether a programme is purely defensive and legal under the BTWC or whether it has quite other aims. Even if the programme is huge and is detected – as the Soviet one was in the West – getting to the detail is much more difficult. It should be no surprise, therefore, to find that the historical record is littered with allegations of biological warfare. It was alleged that the United States used biological weapons against North Korea and China during the Korean War and that it attacked Cuba with those weapons on a number of occasions. Similarly, it has been alleged that the Soviet Union tested biological weapons in Asia by supplying them to sympathizers. Most of these allegations have not stood up well to subsequent analysis, and many are clearly false.

However, one allegation made against the Soviet Union is certainly true, as Russian and American scientists have demonstrated.[6] In April 1979 there was an outbreak of anthrax in the Soviet city of Sverdlovsk (now Yekaterinberg). The number of people who died was less than one hundred and the incident was reported in the émigré press quite quickly. The United States subsequently made the charge that the outbreak was caused by an accidental release from a military microbiology facility. The Soviet Union responded by saying that the outbreak resulted from people eating contaminated meat and not from a release from a covert programme.

After the collapse of the Soviet regime a team of US scientists and social scientists visited the city twice. Two lines of evidence strongly supported the theory of an accidental release of anthrax. Firstly, Russian pathologists had kept some of the post-mortem material rather than let it be collected by the KGB after the outbreak. This allowed a joint reinvestigation of tissue samples and investigation notes from forty-two autopsies. These samples and notes clearly indicated that death had resulted from inhalation anthrax, not gastrointestinal anthrax, as would have been the case if they had eaten contaminated meat. Secondly, the epidemiological evidence showed that at the time of the incident, of the sixty-six fatal cases, sixty-one were in a narrow sector of the city extending away from the military compound and three of the remaining five had occupations (for example, truck driving) that could have taken them there. This distribution could not have happened by chance – particularly since the winds on the day of the incident would have deposited material leaked from the facility in precisely this small section of the city. Thus the victims were in exactly the place where an aerosolized anthrax exposure would have occurred given the prevailing wind conditions of the day.

These innocent Soviet citizens were therefore the unwitting victims of their own country's offensive biological weapons programme. Their deaths should serve as a strong warning to us all of the dangers of "medicine in reverse".

## conclusion

What the last three chapters have shown is that there has been a steady application of the growing understanding of the life sciences in a series of offensive biological weapons programmes carried out by major states over the last century. Furthermore, this history is hardly known in any detail to politicians, the general public or even most scientists today. What the history suggests is that this process will continue in the twenty-first century unless very determined steps are taken to halt it.

## references

1. Office of Technology Assessment (1993) *Proliferation of Weapons of Mass Destruction: Assessing the Risks.* Washington: United States Congress.
2. For data on Iraq's biological programme see Pearson, G.S. (1999) *The UNSCOM Saga.* Basingstoke, England: Palgrave.
3. See Gould, C. and Folb, P. (2003) *Project Coast: Apartheid's Chemical and Biological Weapons Programme.* Geneva: United Nations Institute of Disarmament Research.
4. See Hart, J. (in press) The Soviet BW programme. In M. Wheelis, L. Rozsa and M.R. Dando (eds), *Deadly Cultures: Bioweapons from 1945 to the Present.* Cambridge. MA: Harvard University Press.
5. Alibek, K. and Handleman, S. (1999) *Biohazard: The Chilling True Story of the Largest Covert Biological Weapons Program in the World – Told from the Inside by the Man Who Ran It.* New York: Random House.
6. See Guillemin, J. (1999) *Anthrax: The Investigation of a Deadly Outbreak.* Berkeley: University of California Press.

# biological agents

## introduction

Biological weapons are best regarded as part of a biochemical threat spectrum that ranges from so-called classical (lethal) chemical weapons through poisonous industrial chemicals and mid-range agents such as toxins and bioregulators to traditional biological agents and genetically modified agents (see Figure 5.1). Within this spectrum, biological agents include well-known pathogens like anthrax, toxins produced by bacteria such as botulinum toxin, and normal signalling molecules of living organisms (bioregulators) which, in unusual amounts, can cause massive disruption of normal physiology.

When concerns about possible biological weapon attacks grew in the 1990s, the Centers for Disease Control and Prevention (CDC) in the United States were asked to review the most dangerous threats to the civilian population. They made their judgements using criteria such as:[1]

1. public health impact based on illness and death;
2. delivery potential to large populations based on stability of the agent, ability to mass produce and distribute a virulent agent, and potential for person-to-person transmission of the agent;
3. public perception as related to public fear and potential civil disruption; and
4. special public health preparedness needs based on stockpile requirements [for example, vaccines], enhanced [disease] surveillance, or diagnostic needs.

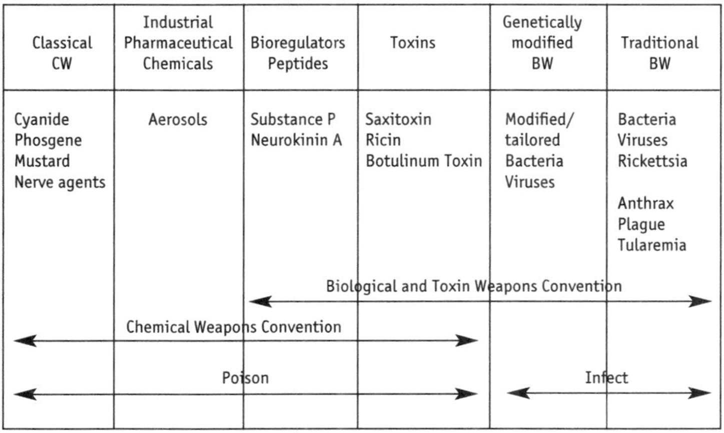


**Figure 5.1** *The biochemical threat spectrum. In an age of molecular biology it is no longer credible to think of separate chemical and biological weapons. It is best to consider the full range of biochemical threat agents as a continuous spectrum as shown here.*

Using these criteria, a list of agents posing the greatest threat to civilian populations was drawn up. The most dangerous of these were designated Category A agents and included such things as smallpox and anthrax (see Table 5.1).

The CDC also produced a list of Category B and C agents which, though not as dangerous as those in Category A, nevertheless

**Table 5.1** CDC Category A agents*

| **Biological agent(s)** | **Disease** |
|---|---|
| *Variola major* | Smallpox |
| *Bacillus anthracis* | Anthrax |
| *Yersinia pestis* | Plague |
| *Clostridium botulinum* (botulinum toxins) | Botulism |
| *Francisella tularensis* | Tularemia |
| Filoviruses and Arenaviruses (e.g. Ebola virus, Lassa virus) | Viral haemorrhagic fevers |

* From Rotz et al., op. cit.

presented a considerable risk. However, when the United States National Institute of Allergy and Infectious Diseases (NIAID) began its programme to develop countermeasures, it used a slightly more developed listing for its research agenda in which the individual agents were organized into various groups (see Table 5.2).[2] The broad grouping of agents in the B and C categories shown in Table 5.2 is much easier to understand than if the individual agents had simply been presented in one long list.

**Table 5.2** NIAID Category B and C priority pathogens*

| Group | Biological agent (examples) | Disease/common name |
|---|---|---|
| *Inhalation Bacteria* | | |
| | *Brucella* species | Brucellosis |
| | *Burkholderia pseudomallei* | Meliodosis |
| | *Burkholderia mallei* | Glanders |
| | *Coxiella burnetii* | Q-fever |
| | *Rickettsia prowazekii* | Typhus |
| | *Rickettsia rickettsii* | Rocky Mountain spotted fever |
| *Arthropod-borne viruses* | | |
| Alphaviruses | Venezuelan equine encephalitis | VEE |
| | Eastern equine encephalitis | EEE |
| | Western equine encephalitis | WEE |
| Flaviviruses | West Nile virus | WNV |
| | Japanese encephalitis virus | JE |
| | Tick-borne encephalitis virus | TBE |
| | Yellow fever virus | YF |
| Bunyaviruses | California encephalitis virus | CE |
| | La Crosse virus | LAC |
| | Crimean–Congo haemorrhagic fever virus | CCHF |
| *Toxins* | | |
| | *Ricinus communis* | Ricin toxin |
| | *Clostridium perfringens* | Epsilon toxin |
| | Staphylococcal enterotoxin B | SEB |

| | | |
|---|---|---|
| *Food- and water-borne pathogens* | | |
| Bacteria | Shiga-toxin-producing *Escherichia coli* | STEC |
| | *Shigella dysenteriae* 1 | Shiga bacillus |
| | *Salmonella typhi* | Typhoid fever |
| Viruses | Hepatitis A virus | HAV |
| | Caliciviruses | e.g. Norwalk |
| Protozoa | *Toxoplasma gondii* | Toxoplasmosis |
| *Emerging infectious diseases* | | |
| | Influenza A virus | Influenza |
| | Multi-drug-resistant tuberculosis | MDR-TB |

* From NIAID, op. cit.

There are clearly many potential pathogens and toxins that could be used in various kinds of biological attacks against people. The essential characteristics of the more important of these agents will be reviewed and some of the lesser-known bioregulators will be considered. At the end of the chapter some of the main agents that could be used to attack staple and economically valuable crops and animal husbandry will also be briefly reviewed. However, in order to understand the problem of biological warfare and terrorism, and how it might best be prevented, it is necessary first to know the general characteristics of the individual agents, be they bacteria, viruses, fungi, toxins, bioregulators or protozoa.

*Bacteria* are single-celled micro-organisms, visible under an ordinary light microscope, in which the hereditary material (DNA) contained within chromosomes is not enclosed within a nuclear membrane. Technically they are therefore called “prokaryotes”. There are vast numbers of different types of bacteria and we have little idea at present just how many different types exist. Bacteria are found everywhere, for example in soil (in enormous numbers), in water, in animals and plants and in the most extreme environments, both hot and cold. Most are harmless, but some bacteria are pathogens that cause important diseases. One example is *Bacillus anthracis*, the cause of anthrax. This living organism is scientifically designated by two names, as are all other living things: *Bacillus* denotes the genus or particular group of bacteria to which it belongs and *anthracis* is the specific name of that particular organism within the group. The whole italicized name, *Bacillus anthracis*, identifies that particular organism to any scientist anywhere in the world, whatever his or her native language. There are, however, many different strains of anthrax, which vary, for example, in their lethality.

*Viruses* are minute infectious agents which can only multiply if they get inside a living cell (their host cell) and subvert its enzymatic machinery. They are not regarded as living because they have no such activity outside the cells they parasitize. They do, however, contain hereditary information in the form of RNA (ribonucleic acid) or DNA (deoxyribonucleic acid) which is transcribed and replicated in the host cell in order to form new virus particles or virions which, when coated with proteins, can burst out of the cell and go on to infect other cells. Viruses are not cellular organisms like bacteria, but they *are* still the cause of major diseases such as smallpox and yellow fever. Though smallpox virus is often called *Variola major*, most viruses are not designated in the same way as living organisms with a binomial name, but are often named for the disease they cause: for example, yellow fever virus causes yellow fever.

*Toxins* are chemical substances produced by living organisms which are poisonous to other organisms. Thus the bacterium *Clostridium botulinum* produces botulinum toxins, which are deadly poisons to human beings. Often, toxins such as the poisons produced by snakes have become exquisitely adjusted through evolutionary time to exert their natural effects. It is therefore difficult to synthesize chemicals that are much more effective than these natural toxins. *Bioregulators*, on the other hand, are natural signalling molecules in living organisms (such as neurotransmitters in the brain) and have not been subject to evolutionary predator/prey interactions. It is therefore often possible to find chemical substances, for example drugs of abuse, that are much more powerful than the natural bioregulators, or to find ways of significantly modifying the operations of natural bioregulators in a way that would not happen in nature. For example, the lethal nerve gases developed during the Second World War (but not used) powerfully disrupted the acetylcholine neurotransmitter system and would have rapidly killed anyone exposed to them.

*Protozoa* are single-celled or colonial animals. They are found everywhere, particularly in aquatic or damp environments. Some are very dangerous, causing human diseases such as malaria, the causative agent of which is *Plasmodium falciparum*. This protozoan is injected into the bloodstream when an infected mosquito feeds on a mammal such as ourselves. Unlike the bacteria, these organisms have their hereditary material contained within a separate membrane-bounded nucleus within the cell.

*Fungi* similarly have hereditary material enclosed in a nucleus, but they do not have distinct cells, each controlled by its own nucleus. Rather they are termed "acellular" organisms. Unlike plants, they lack chlorophyll and so are unable to synthesize their own food by photosynthesis. Instead, they obtain the required nutrients through parasitism or symbiotic means. Fungi are particularly important causes of disease in plants, and several species have been weaponized as biological warfare agents.

With that brief introduction to the biological background, some of the potential biological warfare agents of increasing concern today can be reviewed, beginning with agents that could be used to attack people. Many of these particular organisms are natural pathogens of other animals; they are zoonotic organisms that cause so-called "zoonoses" in the human species.

## anti-personnel BW agents

The organisms listed as Category A and, to some extent also, those in Categories B and C of Tables 5.1 and 5.2 include those which were weaponized in the offensive biological warfare programmes of the last century. It must be appreciated that the agents weaponized were chosen not by chance but because they had certain characteristics useful to an attacker.[3] Some of these characteristics are as follows:

1. An agent should produce a certain effect consistently.
2. The dose needed to produce the effect should be low.
3. There should be a short and predictable incubation period.
4. The target population should have little or no immunity.
5. Treatment for the disease should not be available to the target population.
6. The user should have the means to protect troops and civilians.
7. It should be possible to mass-produce the agent.
8. It should be possible to disseminate the agent effectively.
9. The agent should be stable in storage and transportation in munitions.

Thus, for an attacker, it clearly makes sense if the agent produces its effect at a low dose and if the target population cannot be protected via immunity or medical treatment.

It is also important to realize that an attacker could have any one of a variety of different objectives in carrying out a biological weapons attack. A criminal or terrorist group might use a small amount of agent in an assassination attempt, or a modest amount of agent in order to cause public fear and disorder. A state might use a large-scale release of biological agent for strategic military purposes or as a weapon of mass destruction against a civilian population (Table 5.3). So we are not dealing with just one type of "biological bomb" but a very wide range of possible types and scales of attack.

**Table 5.3** Types of biological attack

| **Scale of agent release** | **Nature of aggressor** | | |
|---|---|---|---|
| | ***Individual*** | ***Subnational group*** | ***State*** |
| Point source | e.g. criminal act | e.g. assassination | e.g. assassination |
| Medium-scale | e.g. criminal act | e.g. terrorism | e.g. military tactical |
| Large-scale | Not possible | e.g. national liberation army (use) | e.g. military strategic |

A further complication arises because not all agents would kill. Some would certainly cause high rates of mortality (such as inhalation anthrax if not quickly treated), but others, some of which were in fact weaponized, would cause incapacitation rather than death in most victims. One such incapacitant is staphylococcal enterotoxin B (SEB), which was extensively studied in the US offensive programme (see Chapter 3). Such an incapacitant might be considered useful, for example, if enemy troops were mixed with civilians or if the intention was not only to disable enemy forces but also to overload their capacity for dealing with the injured.

Finally, it is crucial to differentiate between agents that are not contagious person-to-person after first use and those which are. Agents such as smallpox which are highly contagious in this way are much more difficult for a defender to handle. From an attacker's position, such agents could be expected to have greater effects, but their use brings the potential drawback of creating uncontrollable epidemics.

From a military point of view, therefore, individual agents are viewed against a matrix of possibilities (Table 5.4). Against this wider context, it is possible to see why certain agents are so favoured, in spite of the vast range of available pathogens known to cause illness in humans. We shall deal first with Category A agents and then with some of those in Categories B and C.

**Table 5.4** A military classification of agents*

| **Principal characteristics** | **Examples** | **Militarily significant features** |
|---|---|---|
| *Potentially contagious from first victim* | | |
| Incapacitating | Influenza virus | Limited use because of |
| Lethal | *Yersinia pestis* (plague) | possible lack of control |
| *Not contagious from first victim* | | |
| Incapacitating | *Coxiella burnetii* (Q-fever) | Decay rate in air; incubation |
| Lethal | *Bacillus anthracis* (anthrax) | period; length of illness; etc. |

* From Dando, op. cit.

## category A agents

*Bacillus anthracis*, the causative agent of anthrax, is a rod-shaped, non-motile bacterium large enough to be visible under an ordinary light microscope. As we saw in Chapter 2, the anthrax bacterium is important historically because it was the first micro-organism conclusively shown to be the cause of a disease. In 1876 Robert Koch published his work on the disease. He showed that if he injected material from diseased animals into healthy mice, these mice also contracted the disease. He then incubated the spleen of a dead infected animal in beef broth where the vegetative (growing) bacteria grew, reproduced and eventually produced their characteristic environmentally resistant spores. When the isolated vegetative bacteria or the spores were injected into healthy mice, the mice developed anthrax.

Koch put forward his famous postulates which did much to underpin the rapid discovery of many more bacterial agents of disease at the end of the nineteenth century. The postulates were:

1. The micro-organism must be present in every case of the disease but absent from healthy animals.
2. The suspected micro-organism must be isolated and grown in pure culture.
3. The disease must result when the isolated micro-organism is inoculated into a healthy host.
4. The same micro-organism must be isolated again from the diseased host.

Having used this scientific approach in his study of anthrax, Koch went on to discover the organisms that cause tuberculosis and cholera. On the basis of such scientific analysis our knowledge of pathogenic micro-organisms has since been developed.

### *anthrax*

Anthrax is a natural pathogen of grazing mammals. Until the development of effective vaccines against the disease for these animals it was a major cause of fatal illness and it continues to kill domestic and wild animals where vaccination programmes are not maintained, for example during the disruption caused by warfare. The environmentally resistant spore form of the organism is found in the ground where infected animals have previously died. Grazing animals can therefore ingest the spores or inhale them. Once inside an animal's body the vegetative form of *Bacillus anthracis* develops, grows, multiplies and produces several toxins which together kill the infected animal. When the animal dies and its body decays, nutrients run out, oxygen becomes freely available and the environmentally resistant spores form again and can remain for long periods in the soil, ready to infect other animals. With these characteristics, it is not surprising that in the past anthrax was also known as "woolsorter's disease" in England, since it was a hazard for workers involved in processing fleeces and wool.

Humans can contract three different forms of anthrax disease. If the spores get into a skin cut then cutaneous anthrax can develop. Eating infected meat can result in gastrointestinal anthrax. In a bioweapons attack, however, the aim would be to spread the spores on the air so that they were taken into the lungs to cause inhalation anthrax. This disease develops over a period of one to seven days and is deadly if untreated. The disease begins with non-specific, influenza-like symptoms that make early diagnosis difficult, but rapidly worsens with many dangerous symptoms appearing. Death rates for untreated cases are at least ninety per cent.

Anthrax is not contagious person-to-person, so there is no need to quarantine those infected. Various means can be used in the laboratory to diagnose the presence of *Bacillus anthracis* in the body and antimicrobial therapy is effective if begun early enough. A vaccine is available in the West for people at risk of contracting the disease, but it requires a series of injections over time. It is uncertain whether the vaccine would be effective against a heavy aerosol inhalation assault. The dose needed to infect an individual is also the subject of great uncertainty. The standard measure used is the dose required to infect fifty per cent of a population and this, the LD50, can only be estimated from animal experiments.

Anthrax has been a standard choice for those interested in causing disease in animals and people. It was used in the German anti-animal campaigns of the First World War and was weaponized as a retaliatory anti-animal weapon by the British in the Second World War. It was later weaponized as an anti-personnel agent by both sides in the twentieth-century East–West cold war (see Figure 3.1 and Chapter 4). More recently, anthrax has come to public attention following the deaths caused by leakage of the agent from a Soviet military facility in Sverdlovsk in 1979 (Chapter 4) and the anthrax letter attacks in the United States in late 2001.

As will become apparent, preparing an agent like *Bacillus anthracis* for use in an airborne biological weapons attack is not an easy technical task. It appears to have taken years of experimentation to perfect in the state offensive biological weapons programmes of the last century. Such "weapons grade" material must have had a very high concentration of spores, uniform particle size and low electrostatic charge and been subject to other treatment to reduce clumping of the material. There is, however, a prior problem for the would-be attacker. Because of the natural processes of mutation and geographical isolation in different environments, there are many different strains of anthrax around the world and these have different levels of lethality in humans. The first problem for a weaponeer, therefore, would be to find a virulent strain of anthrax.

It is not a well-known fact that, before their unfortunately successful attack on the Tokyo underground with sarin nerve gas in 1995, the Aum Shinrikyo sect had attempted to use anthrax to attack their fellow citizens. It has been suggested that one of the reasons they failed in their biological weapon attacks was that they did not have access to a lethal strain of the organism. The known difficulty of preparing weapons grade anthrax has led many commentators to surmise that the anthrax letters sent in the US in 2001 must have used material (or had inside information) from US official sources.

### *smallpox*

There are two kinds of smallpox. That caused by the virus *Variola major* commonly kills 25–30 per cent of those infected whereas infection caused by the vaccine strain *Variola minor* has a death rate of one per cent or less. Closely related diseases such as cowpox and monkeypox exist, but smallpox appears to be uniquely an infection of humans. Smallpox is a DNA virus (its hereditary material being DNA, not RNA) and a member of the orthopoxvirus genus of viruses. The individual organisms which are called virions, are among the largest and most complex of all viruses. The virion has a brick-shaped structure with a maximum dimension of about two hundred nanometres and is therefore easily visible under an electron microscope.

Smallpox caused by *Variola major* is an acute viral disease usually contracted by breathing in airborne droplets that carry the virus. In the past it seems likely that as many as ninety per cent of people exposed contracted the disease. It affected people of all races and did not discriminate between young and old or male and female. The infection had only two possible outcomes – death or long-lasting immunity. With no known animal reservoir, the disease could only exist as an active infection, causing waves of epidemic disease at different times and places in history.

With these characteristics, it seems unlikely that the disease could have existed in the sparse populations of early human history. However, it is possible that it was present in ancient Egypt. As is evident from his mummified remains, Pharaoh Ramses V, who died in 1157 BCE, had a face, neck and shoulders disfigured by a rash of elevated pustules that could have been caused by smallpox. Epidemics that might well have been smallpox also raged across the Roman Empire in the second and third centuries CE, but the descriptions of the victims are not clear enough to be sure. By the ninth century CE, though, there are clear descriptions by a physician in Baghdad which leave no doubt that he was differentiating between measles and smallpox.

The populations of the Old World, in Europe and the Middle East, were accustomed to smallpox as a disease by the sixteenth century. But as these peoples began to expand and migrate to the New World and Asia, they encountered populations that had had no contact with the disease. In those circumstances, its effects could be devastating and, as we saw in Chapter 2, Europeans knew enough about the disease – even though they did not understand it scientifically – to use it for hostile purposes against the North American Indians.

If the parish records of burials in England in the eighteenth century are examined, it is also apparent that smallpox was one of the few diseases identified as a cause of death. Our modern techniques of vaccination had their origin at this time in efforts to prevent smallpox. The practice of "variolation" arose in which people were deliberately infected by obtaining material from an active case and scratching it into the skin. Practised people who did this expertly were said to be able to keep the disease in a mild form with a very low consequent death rate. Edward Jenner in England noticed that variolation failed to produce the symptoms of smallpox in people who had previously suffered the mild disease cowpox. He then vaccinated people with cowpox and showed that later variolation reliably failed to produce the smallpox disease. Jenner published his findings in 1798 and vaccination was rapidly taken up around the world. Eventually, a campaign organized by the World Health Organization eradicated the disease in 1979.

After the eradication campaign it was intended that stocks of the virus be kept in secure storage in just two places – the United States and the Soviet Union. Unfortunately, if the accounts of some of those involved are to be believed, as part of its illegal offensive biological weapons programme the Soviet Union weaponized massive amounts of smallpox as a lethal anti-personnel agent. We can only hope that all such material has now been destroyed and that the exercise will never again be repeated. Since very few people now have effective protection from vaccination against smallpox (having been vaccinated long ago or never having been vaccinated at all), an outbreak of this very contagious lethal disease could be devastating.

The infectious dose for smallpox is not known, but it is thought to be just a few virions lodged in the oropharyngeal or respiratory mucosa. At the end of an incubation period of 7–17 days, the patient experiences a high fever and is prostrated with headache and backache. A rash then appears which, within a couple of days, turns vesicular and then pustular. The pustules are deep in the skin and leave pitted scars if the patient survives. The patient is most infectious to others for 7–10 days after the rash appears. Although ninety per cent of cases follow this characteristic pattern, there are two other forms of the disease. In haemorrhagic smallpox there is widespread haemorrhaging into the skin and mucous membranes and the patient invariably dies about five days after formation of the rash. In the frequently fatal malignant variant, the pustules do not appear and the skin takes on a reddish-coloured rubbery form. Diagnoses of smallpox in its characteristic form were made from the nature and distribution of the rash. Haemorrhagic and malignant smallpox were much more difficult to diagnose.

Just one case of smallpox nowadays would be the cause of a worldwide emergency. At present, all the treatment available to someone with smallpox would be supportive care and antibiotics to prevent secondary infection. There are currently no antiviral agents that are effective against smallpox, but vaccination administered within a few days of exposure might prevent or ameliorate the disease. A major problem arises from the known difficulty of preventing smallpox transmission in hospitals. Not only is the virus capable of airborne infection, perhaps for twenty-four hours, but it can remain viable on laundry from infected people for extended periods of time (see the account of the early hostile use of smallpox in Chapter 2). Furthermore, though many people were safely vaccinated in the past, there was, and is, always the danger of major complications for a few.

### *plague*

The non-motile, non-spore-forming bacterium *Yersinia pestis* is the cause of plague. The bacterium is easily visible under a light microscope, where stained preparations show a characteristic bipolar or "safety pin" appearance. Though it can remain viable for some days in water or moist soil the bacterium is killed within a few hours by direct sunlight. Plague is still a problem disease in some parts of the world: there were outbreaks of plague in humans in Africa, Asia and South America during the 1990s.

*Yersinia pestis* is a natural pathogen of rodents such as the black rat, *Rattus rattus*, and the brown rat, *Rattus norvegicus*. The pathogen is transmitted between rodents and other animals through the bites of infected fleas. Major outbreaks of plague can occur in cities when many rats are infected and the disease spreads to humans. The normal human disease is bubonic plague, caused when the pathogen enters the body via regurgitation during a flea bite or through broken skin. If the lungs become infected, then a much more deadly, pneumonic, form of plague results. It is this pneumonic form that would result from an aerosol attack if the agent was used in a biological weapon. Because there are natural reservoirs of the disease in rodent populations around the world, because it can be mass produced and disseminated, because there is a high fatality rate in untreated cases and because pneumonic plague is contagious through airborne spread, it is readily apparent why there is great concern about the possible use of *Yersinia pestis* as a biological weapons agent.

The "Black Death" is the name given to the great pandemic of plague which occurred in Asia, the Middle East, North Africa and Europe in the middle of the fourteenth century. The pandemic seems to have originated in 1346 north and west of the Caspian Sea. From there it reached Constantinople in 1347, then moved south and east through the eastern Mediterranean and the Middle East and west and north through the western Mediterranean and Europe. By 1348 it had arrived in southern England and in 1353 it was in Moscow. So the height of the long-running disease pandemic lasted about seven years, sweeping along in waves that consumed the peoples of the regions affected. The duration of the pandemic in any one place appears to have been about six months. With the centuries that have elapsed, it is difficult today to judge the death rates during the pandemic, but historians generally agree that it was in the range of 30–50 per cent of the population. The effects of this mortality rate on a society that had not suffered such an outbreak for centuries were enormous. The available control measures were largely ineffective against a disease whose cause was unknown, and there was considerable social disruption and change as a result of the pandemic. The Black Death was, in fact, the peak of a second pandemic of plague. The first pandemic occurred in the mid-sixth-century CE and the third began in China in 1855, killing some twelve million people in India and China alone.

Infected people develop symptoms of bubonic plague between two and eight days after being bitten by an infected flea. They experience sudden fever and weakness and the characteristic bubos (swollen, tender lymph nodes) appear about a day later. A small number of people do not develop bubos but instead have primary septicaemic plague. Septicaemia can also develop following the appearance of the bubos. The name "Black Death" may have arisen from the appearance of gangrene in the nose, digits and other extremities following the onset of septicaemia. Secondary pneumonic plague can also develop in a minority of people suffering from bubonic plague. As mentioned earlier, following an aerosolized biological attack people would exhibit primary pneumonic plague and there would be no tell-tale bubos to aid diagnosis or, unfortunately, any widely available, rapid diagnostic tests.

Until 1999 a licensed vaccine was available in the United States, but its production has been discontinued and, in any case, it apparently did not prevent or ameliorate the development of primary pneumonic plague. This is probably true of other vaccines available around the world. Vaccination during an epidemic would not be of much help anyway, since immunity takes a month to build up. It *is* possible to treat people suffering from plague successfully with antibiotics, but if this treatment is not begun within twenty-four hours of the onset of symptoms the fatality rate remains very high.

Under modern conditions the outbreak of a new plague pandemic seems improbable. It should not be forgotten, however, that Japan tried to use plague-infected fleas to cause an outbreak among the Chinese during Japan's mid-twentieth-century offensive biological weapons programme (see Chapter 2) or that the Soviet Union later succeeded in mass-producing *Yersinia pestis*. Furthermore, it is clear that antibiotic resistance can be built into this organism using modern techniques of genetic engineering. Plague well deserves its place in the Category A list of potential biological threat agents.

### *botulinum toxin*

Botulism is caused by the extremely potent toxin produced by the bacterium *Clostridium botulinum*. Botulinum toxin is the most poisonous substance known to humankind. It has been estimated that, if evenly dispersed and inhaled, one gram could, theoretically, kill more than a million people. *Clostridium botulinum* is a spore-forming bacterium whose spores are found naturally in the soil. The toxin is produced by the growing (vegetative) form of the bacterium when it is in oxygen-limited environments such as a wound or in canned food that has not been properly sterilized to kill the organism.

Seven distinct antigenic types of the toxin (A–G) have been recognized. This means that an antitoxin to the A type does not neutralize the B–G types and so on. Botulism in humans is generally caused by the A, B, E or F toxins. Types C and D have been shown to cause disease in other animal and bird species. The toxin is not absorbed through intact human skin but can enter through a wound or through mucosal surfaces such as in the gut or lung. Distressing signs of the disease then follow rapidly within 12–72 hours of infection, say through eating contaminated food.

People suffering from botulism experience dysfunction of their motor nerves, so, for example, they may have blurred vision and difficulty in speaking or swallowing. If not treated rapidly and effectively, they eventually die because general muscular paralysis also affects their respiratory muscles and they are unable to breathe. In the twentieth century botulinum toxin was soon recognized to be a potential lethal biological weapons agent. The Japanese tested its effects on prisoners in their offensive biological weapons programme, and because of concerns that Germany might use it against the Allies on D-Day more than one million doses of botulinum toxoid vaccine were prepared. Curiously, the toxin has recently been licensed for treating certain medical conditions such as types of muscle spasm, and for cosmetic reasons to remove wrinkles.

The mechanism of action of the toxin has been worked out in considerable detail. Human motor neurons transmit information to the muscles they supply in the following way: a nerve impulse travelling down the axon of the neuron causes the release of acetylcholine neurotransmitter where the nerve ends on muscle, and the muscle responds by contracting. The chemical acetylcholine is stored in vesicles at the nerve ending and when the nerve impulse arrives it normally causes them to fuse with the neuronal membrane and release their contents onto the muscle. When botulinum toxin gets into a neuron it prevents the vesicles fusing with the outer membrane of the neuron and thereby prevents release of acetylcholine. There is no input to the muscle and the muscle does not contract.

The major concern is that aerosolized botulinum toxin might be used to cause widespread inhalation botulism, but contamination of the food supply could also result in a large number of cases for the medical services to deal with. Botulism is, unfortunately, easily confused with other diseases of the nervous system, and laboratory tests that take days to complete are needed to confirm diagnosis. Modern medical care can greatly diminish the death rates from botulism;

antitoxins are available and, if administered early, can limit the damage to nerves. However, if damage does occur, the patients may require prolonged treatment, including feeding and mechanical ventilation, until their nervous systems recover. In theory, it would be possible to eliminate the hazard of botulinum toxin by mass immunization, but this is unlikely to happen, partly because of the scarcity of the required toxoid.

### *tularemia*

The disease tularemia, caused by the bacterium *Francisella tularensis*, is much less known to the general public than the other potential biological weapons agents in Category A. Yet it is an agent that was investigated by the Japanese and by both the United States (Chapter 3) and the Soviet Union (Chapter 4) in the last century. The bacterium normally infects various wild animals and only causes disease in humans occasionally, say through an insect bite when there is a large natural outbreak or someone is out in wild hunting country. For that reason the disease is known in different parts of the world as an animal fever, for example "rabbit fever".

Tularemia was first described in 1911 as a potentially severe and fatal disease for humans. Large-scale human epidemics occurred in the 1930s and 1940s in Europe and in the Soviet Union, and the bacterium began to be recognized as a significant laboratory hazard for those working with it because of its extreme infectivity. There are two predominant sub-types of the organism, sub-type A being much more virulent than sub-type B. We know now that the organism is one of the most infectious of the pathogenic bacteria, perhaps as few as ten organisms being sufficient, if inhaled, to cause the disease in a human. Person-to-person transmission of the disease has not been documented. Although the bacterium does not form a spore, it can survive for weeks in moist soil, hay, straw and the like if temperatures are low.

Clinical manifestations of the disease vary with the route of entry and the virulence of the organism. Infection through the skin, for example, usually produces an ulceroglandular form of the disease, with an ulcer at the point of entry and swelling of the local lymph nodes. There is also an abrupt onset of fever, malaise and joint and muscle pains. Infection via an aerosol could produce a variety of symptoms such as pharyngitis or it might just be manifest as a systemic illness without such signs. An outbreak of this kind in a population would result from a successful biological weapons attack after about 3–5 days and would be very difficult to distinguish initially from an outbreak of influenza or an attack using a variety of other agents (see, for example, Q-fever below).

Vaccines are available to prevent tularemia infections and have been widely used in Russia since the 1930s. The development of better ones is hindered by the lack of detailed knowledge of the pathogenesis caused by the microbe. Antibiotics are effective against tularemia, but rapid confirmatory diagnosis is not simple. A source of much concern is that in both the Soviet and US offensive programmes antibiotic-resistant forms of the organism were studied. The overall mortality rate from the more virulent A strain is potentially as high as sixty per cent of untreated cases, so there is every reason for anxiety about this potential agent.

### *viral haemorrhagic fevers*

The term "viral haemorrhagic fever" (VHF) refers to diseases that produce fever and bleeding as a result of infection by viruses from one of four different families (Table 5.5). The viral haemorrhagic fever viruses are all small and their hereditary material is RNA. The viruses are transmitted to people through contact with infected animals or by arthropod vectors. The course of the disease varies with each different virus, but there is still a great deal to be learned about their natural history and the pathogenesis they cause in humans.

In the early 1990s the *Cambridge World History of Human Disease*, in its section on Ebola virus disease, stated, "Textbooks on tropical diseases in Africa are well out of date. With the recognition

**Table 5.5** Some haemorrhagic fever viruses*

| Family | Virus | Disease |
|---|---|---|
| Filoviridae | Ebola | Ebola haemorrhagic fever |
| | Marburg | Marburg haemorrhagic fever |
| Arenaviridae | Lassa | Lassa fever |
| Bunyaviridae | Rift Valley fever | Rift Valley fever |
| Flaviviridae | Dengue | Dengue fever |
| | Yellow fever | Yellow fever |

* From Rotz et al., op. cit.

of new and deadly viral infections – Lassa, Marburg, Ebola ... the classical descriptions of major diseases ... must be thoroughly revised." As the authors noted, major discoveries had recently been made. They continued, "After the appearance of the Marburg virus in 1967 and the Lassa virus in 1969 had given a jolt to complacency, the Ebola virus in 1976 provided a compulsive shudder." Of the 284 detected cases in the June 1976 outbreak of Ebola virus in the Sudan, 148 died – a fifty-two per cent mortality rate. In the subsequent September outbreak in Zaire there were 288 deaths out of the 318 cases – 90.5 per cent mortality. A smaller outbreak in Zaire in 1979 resulted in sixty-six per cent mortality. There were no substantial reported outbreaks through the 1980s until, in 1989, a shipload of monkeys from the Philippines, destined for the United States, was found to be infected. Sixty of the one hundred monkeys died but fortunately the outbreak was contained.

If the VHF viruses are considered as potential biological weapons agents, it is clear that some are not suitable. Dengue, for example, is not transmissible as a small-particle aerosol. Others are the cause of considerable concern. In the Soviet Union's offensive biological weapons programme, it was demonstrated that just a few virions of Marburg virus were sufficient to cause infection, and large quantities of this virus, along with Ebola and Lassa viruses, were produced. The United States also investigated yellow fever and Rift Valley fever viruses in its offensive programme. Rift Valley fever and the flaviviruses (yellow fever etc.) are not transmissible person-to-person, but agents like Ebola virus can spread through close contact very effectively if special precautions are not taken.

There is an effective vaccine for people travelling to areas where yellow fever may be present, but there is no licensed vaccine, even in the United States, against any other virus in the VHF group. The yellow fever vaccine, moreover, could not be used *after* an attack because the disease has a shorter incubation period than the time needed for the development of antibodies. In any case, the vaccine is in relatively short supply worldwide.

Filoviruses like Ebola and Marburg are extremely virulent in non-human primates and in humans: infection results in widespread damage to the viscera (such as the liver, spleen and kidneys). The variable clinical picture that can result from infections with these viruses could make diagnoses very difficult. Confirmatory laboratory tests are available but cannot be completed within hours. The antiviral drug ribavirin may help in some cases, but the main treatment available is careful supportive medicine – which may not be possible if the number of victims of an attack is large.

Studies have shown that viruses like Ebola, Marburg and Lassa can successfully cause infection in non-human primates when prepared in aerosols and inhaled into the lungs, so there is every reason to believe that they could cause massive human casualties in a successful BW attack. Little wonder then that they are placed in the most dangerous – Category A – list of potential biological weapons agents.

## category B and C agents

As evidenced by the groupings of potential agents in the United States NIAID listing of Category B and C agents (see Table 5.2), a wide range of different kinds of attack were considered. The groups – inhalation bacteria, arthropod-borne viruses, toxins, food- and water-borne pathogens and even emerging infectious diseases – show just how diverse the mechanisms of attack could be. It must be clearly grasped that in no sense can the threat from such agents be ignored. A number of these pathogens were weaponized in the offensive biological weapons programmes of the last century. It is not possible to review in detail all the agents listed in Table 5.2, so some illustrative examples are presented here.

### *brucellosis*

In humans this disease results from infection by any of the four main species of *Brucella*. These bacterial species are normal pathogens of animals: *Brucella melitensis* infects goats, *Brucella abortus* cattle, *Brucella canis* dogs and *Brucella suis* pigs. The cause of the disease was first worked out by David Bruce in Malta in 1887. He showed that the undulant (or Malta) fever prevalent among civilians and British troops on the island resulted from infection by *Brucella melitensis* originating from the island's goats. The bacterial cells are able to persist in the environment for weeks and human infections often result from eating raw animal products or drinking unpasteurized milk. In dried preparations the bacteria can remain virulent for years and infection by aerosol requires only a few organisms. Thus infections are common among laboratory staff working with the organism, although person-to-person transmission is rare.

The incubation period for a brucellosis infection is highly variable, usually between five and sixty days. Severe exposure would result in a shorter incubation period. Symptoms include an undulating fever, exhaustion, back and leg pains, headaches and so on. Without treatment, people usually recover after two to three months, but there can be cycles of remission and relapse over years and serious complications can result. Fatality rates are three per cent or less, but the illness is debilitating and, though antibiotic treatment can be successful if begun early enough, no human vaccine is available to protect against the disease. With such characteristics, it is not surprising to find that *Brucella suis* was weaponized as an incapacitating agent in the US offensive biological weapons programme of the mid-twentieth-century.

### *Q-fever*

Another pathogen weaponized as an incapacitant in the US programme was *Coxiella burnetii.* This causes so-called Q-fever in humans. The "Q" stands for "Query" because of the initially uncertain nature of the disease. It was first identified in Australia, being initially recognized as an infection affecting abattoir workers in Brisbane. The scientific name of the organism honours Cox and Burnet, the scientists who made significant discoveries in the early work on the organism.

The pathogen is found in many wild animals besides livestock and its natural life cycle includes transition through tick species. However, it is so infectious to humans in an aerosol and so stable in the environment that infection is usually through inhalation of dust containing the organism. The clinical features of the disease begin after an incubation period of 18–21 days (again a large dose leads to a shorter incubation) and include chills, fever, headache and muscle and chest pains. There can also be nausea, vomiting, diarrhoea and so on. The mortality rate is less than one per cent, but the illness can persist for months. A vaccine is available in Australia and antibiotics can be successful if given early in the infection.

During the US offensive programme in the 1950s, members of the Seventh Day Adventist Church who did not wish to be conscripted into the armed forces could volunteer instead to be subjects for human testing of some biological warfare agents in what was called "Project Whitecoat". One of the agents studied in this way was *Coxiella burnetii,* so there are good data on the infectivity of aerosols of this particular agent and every reason to believe that it would severely affect the health of a large percentage of those exposed to an attack. This agent fulfils very well the requirements laid down for an effective agent: it consistently causes disease at low levels of the organism; it can be manufactured on a large scale; it remains stable during production, storage and transport; it can be effectively disseminated (in an aerosol or as a contaminant of the food supply); and it remains viable for years. So, though it is not contagious and does not cause the high mortality of the Category A agents, Q-fever is still a disease that could cause massive problems for a military organization or a civilian population; its causative agent well deserves its place in the Category B listing.

### *venezuelan equine encephalitis*

When the US offensive biological weapons programme was in operation, much more was known about bacteria than about viruses and so there was much more work on bacterial rather than viral agents. During the later cold war period we know that in the Soviet programme more attention was paid to viral agents. However, one viral agent, Venezuelan equine encephalitis virus, *was* weaponized by the United States. This virus is on the NIAID Category B and C list in the arthropod-borne viruses group (Figure 5.2). The virus is classified as an Alphavirus and the listing also includes the related eastern and western equine encephalitis viruses.

Epidemics of VEE were first recognized in the 1930s and the disease is endemic in the central and northern parts of South America. The virus usually exists through a rodent–mosquito–rodent cycle, but humans become infected naturally through the bite of an infected mosquito when the disease spreads to an equine–mosquito cycle. There is no evidence of direct equine-to-human or human-to-human transmission. However, humans can be infected through an aerosolized agent and again it is clear that very few organisms are required to initiate the disease.

The disease manifests itself as an abrupt onset of influenza-like symptoms: severe headache, high fever, chills and muscle pains. It can also produce nausea, vomiting and diarrhoea. Most infections last 3–5 days and mortality rates are below one per cent. However, there can be major effects on the central nervous system, with severe consequences for a small number of people. Fortunately, there are a number of vaccines available to prevent the disease in humans in affected regions of the world.

### staphylococcus aureus *enterotoxin type B*

Although it was difficult to produce large amounts of many bacterial toxins at the time, the United States, in its offensive biological weapons programme, did weaponize an incapacitating toxin in addition to the lethal botulinum toxin. This incapacitating toxin was *Staphylococcus aureus* enterotoxin type B, or "SEB" as it became known. The five types of staphylococcal enterotoxins are a frequent cause of food poisoning through improperly stored or cooked foods such as ham, processed meats, ice cream and the like. Symptoms are usually nausea, vomiting and diarrhoea and normally occur within 1–6 hours of eating contaminated food. They are of relatively short duration – about twenty-four hours.

Inhalation of staphylococcal enterotoxin B causes a sudden onset of fever, headache, chills and a non-productive cough within a few hours. The victim is likely to suffer these symptoms and be prostrated for up to five days and the cough can persist for weeks. The disabling dose for humans has been estimated to be very small in relation to body weight and the lethal dose to be at least fifty times greater. The toxin is known to trigger the release of massive amounts of cytokines (bioregulators) in the body, which then produce the symptoms in the victim. It is therefore obvious why SEB is on the list of Category B agents of concern.

### *BZ*

Another more direct interference with the body's chemical signalling system is caused by BZ – 3-quinuclidinyl benzilate. This is usually classed as a chemical incapacitating agent. Appropriately, since it was weaponized by the United States during the cold war, it is on one of the schedules of particularly dangerous chemicals subject to special international oversight in the Chemical Weapons Convention. BZ is one of a family of chemicals called glycollates. The Iraqis were said to have produced another, called Agent 15. Such so-called bioregulators are also covered by the prohibition on toxins in the Biological and Toxin Weapons Convention.

BZ interferes with the transmission of information between nerve fibres via acetylcholine in a different way from that already discussed for botulinum toxin. At particular types of acetylcholine transmitter synapses (junctions between the fibres), BZ blocks the receptor on the post-synaptic nerve fibre so that the synapse does not function and the signal is not transmitted further. As will be seen in later chapters, there has been considerable anxiety about the possibility that modern genetic engineering techniques might be used to insert the genes for production of disruptive bioregulators into pathogens, thus giving them alarming, and previously unknown, properties.

### *ricin*

Another toxin on the Category B list is ricin, a highly toxic glycoprotein that is found in the seeds (beans) of the widely cultivated castor oil plant, *Ricinus communis*. Ricin can be dangerous if inhaled or ingested but does not cross intact skin. It acts by inhibiting protein synthesis in the body's cells and causes damage by killing the cells. At present the only treatment is supportive care and there is no vaccine available for human use. Ricin is a danger because it can be extracted relatively easily from castor oil beans. About one million tons of these beans are processed every year and ricin accounts for some five per cent of the residual waste. It is little wonder, therefore, that some of the suspected terrorists arrested recently in Europe were alleged to have been trying to produce ricin, since it is perhaps the easiest such agent to obtain in quantity.

### *a cautionary tale*

The importance of considering attacks other than with aerosolized agents is shown by events in Oregon in 1984. There, on 17 September, the public health department began to be notified of people falling ill with gastroenteritis after eating at restaurants in the small town of The Dalles. Eventually, because people reported eating in salad bars, all salad bars were closed down on 25 September. It was eventually shown that at least 750 people had become ill with salmonellosis, caused by the bacterium *Salmonella typhimurium*. There are many thousands of such cases annually, caused by food contamination with the bacterium, so nothing too unusual was suspected at the time. However, it was later discovered that the salad bars had been deliberately contaminated by followers of Bhagwan Shree Rajneesh.

The sect had purchased a large ranch in the region and was seeking to prevent people voting in a local election in order that the sect could more easily gain permission for developments they wished to make on their land. Commune members were, in fact, just trialling a plan for making people ill on the forthcoming election day for the county commissioners in November 1984. The sect had grown the bacteria in secret laboratories and then poured them onto items in the salad bars. Eventually, the FBI discovered vials containing the bacteria on the sect's ranch and two members pleaded guilty and served prison sentences for their activities.

Though *Salmonella typhimurium* itself is *not* on the list of agents of concern, this criminal activity shows just why food- and water-borne pathogens such as *Salmonella typhi*, the cause of typhoid fever, and *Shigella dysenteriae*, the cause of bacillary dysentery, certainly are on NIAID's Category B and C priority list. If the sect members had carried out widespread food contamination with a more virulent agent in a town, they could have caused many, many more people to become ill – and they might well not have been detected as the source of the contamination.

## anti-agriculture biological warfare

The diversity of the anti-personnel types of biological warfare agent should remind us that this form of warfare need not only be directed against people. Microbial pathogens cause enormous problems in agriculture and some of these pathogens are also suitable for deliberate use. Animal husbandry is particularly vulnerable, in part because it is often very intensive, with many animals kept in confined areas. It is also vulnerable because the animals reared are often from very limited genetic stock, so that a large percentage of them could succumb to a single strain of pathogen. Finally, as is well known from disease outbreaks such as the recent foot-and-mouth disease (FMD) epidemic in the UK, the viral agents that cause disease in animal stocks are often particularly virulent.

### foot-and-mouth disease

This is a very contagious disease of cloven-hoofed animals (cattle, pigs, sheep, goats, etc.). The cause is the FMD virus of the Aphthovirus genus of the Picomaviridae family. There are seven different serotypes of the virus and no cross-immunity between these types. The disease *can* be highly lethal to young calves, but usually lethality is low. The problems are the serious production losses caused by the disease and the measures that have to be taken to eradicate the outbreak.

Natural infections have an incubation period of 2–8 days. The symptoms of FMD are fever, loss of appetite and cessation of milk production in cows. Vesicles develop, particularly around the mouth and the feet, and then rupture to leave painful ulcers. Foot-and-mouth disease can be difficult to distinguish from a number of other infections, and collection and analysis of samples has to be done with great care because the organism is so contagious. Infected animals produce viruses in saliva, milk, faeces, urine and exhaled air. Furthermore, the virus is hardy and is known to have survived kilometres of airborne transmission to cause a new outbreak of disease elsewhere.

## newcastle disease

This is another highly contagious viral disease and affects both domestic and wild birds. The virus belongs to the Paramyxovirus genus and the viruses in that genus affecting birds belong to nine serogroups. Newcastle disease virus is termed A/PMV1. The different strains of the disease vary widely in virulence but some cause high lethality in domestic chickens, turkeys and pheasants.

The incubation period is generally about five days. The effects of the disease are variable, but in its most virulent – viscerotropic velogenic (VVND) – form there is a sudden arrival and spread of the disease. Birds lose appetite and egg production drops off sharply. Profuse bright green diarrhoea is common and the birds suffer extremely rapid dehydration. Many birds die within a day or two and the mortality rate can be over ninety per cent. As with FMD, the virus is hardy and is excreted in faeces and in exhaled air. In affected areas significant reservoirs of infection can take hold in wild birds, providing means for the disease outbreak to be reinitiated.

With characteristics like these, it is little wonder that animal viruses useful for anti-animal biological warfare were investigated thoroughly in the major offensive programmes of the last century. Biological warfare in the First World War was directed by both sides against the valuable draught animal stocks of the other (see Chapter 2). The first really viable biological weapon was produced by the British in their anthrax-laced cattle cakes for potential use against the German cattle industry. Of some relevance today, with worries about possible terrorism, is the fact that pathogens like FMD and Newcastle disease do not affect humans. The people producing and using such agents therefore would be at little risk of infection in the process. Given the importance of animal husbandry, such means could be very attractive to those wanting to damage a country economically. Just imagine the impact of a major, deliberately caused FMD outbreak in a country with a large beef export industry!

## plant pathogens

We just have to be reminded of the nineteenth-century Irish potato famine to realize how devastating fungal diseases can be to staple crops. All staple and economically important crops have to be constantly guarded against the ravages of pests and diseases and even then there can be huge production losses. It is therefore hardly surprising that nearly all state-level offensive biological weapons programmes of which we have knowledge carefully investigated anti-plant attacks.

In 1997, during the negotiations in Geneva designed to strengthen the BTWC (negotiations that later failed), the South African delegation put forward a document[4] that discussed plant pathogens according to the following set of criteria:

1. agents known to have been developed, produced or used as weapons;
2. agents that have severe socio-economic and/or significant adverse human health impacts, owing to their effects on staple crops, to be evaluated against a combination of the following considerations:

- ease of dissemination (e.g. wind, insect, water, etc.)
- short incubation period and/or difficult to diagnose/identify at an early stage
- ease of production
- stability in the environment
- lack of availability of cost-effective protection/treatment
- low infective dose
- high infectivity
- short life cycle

Ten plant pathogens were then identified as potential anti-plant biological weapons agents (Table 5.6).

The potential of plant pathogens was not lost on the bioweaponeers of the last century. The United States, for example, is known to have weaponized agents to attack wheat and rice staple crops (Figure 5.2). As can be seen from the comments in Table 5.6, such attacks on wheat and rice could be extremely destructive. The danger of attacks on crops has increased in recent years because of great advances in our understanding of biocontrol of plant pests and plant inoculants. Indeed, it has been suggested that efforts to develop fungal agents to attack drug crops such as poppies could dangerously contribute to

**Table 5.6** Potential anti-plant biological weapons agents*

| Disease | Agent | Comment |
|---|---|---|
| Coffee berry disease | *Colletotrichium coffeanum* | Could cause serious economic damage |
| Blight of pines | *Dathistroma pini* | Could cause economic damage |
| Fire blight of apple, pear and related species | *Erwinia amyorora* | Could cause economic damage |
| Potato, tomato wilt, Moko disease of banana, etc. | *Pseudomona solanaceorum* | Could cause economic damage |
| Blast disease of rice | *Pyricularia oryzae* | Extremely destructive of this staple crop |
| Maize smut | *Ustilago maydis* | Could cause economic damage |
| Leaf scald of sugarcane | *Xanthomonas albilineans* | Could cause economic damage |
| Bacterial blight of rice | *Xanthomonas campestris* | Extremely destructive of this staple crop |
| Cover smut, stinking smut, common bunt of wheat | *Tilletia tritici* | Extremely destructive of this staple crop |
| Cottony soft rot, white mould and watery soft rot on vegetables, beans, soya, etc. | *Sclerotinia sclerotonium* | Could cause economic damage |

* From South Africa, op. cit.

developing techniques that could be used in biological warfare. As with animal husbandry, the plant species used in agriculture are often

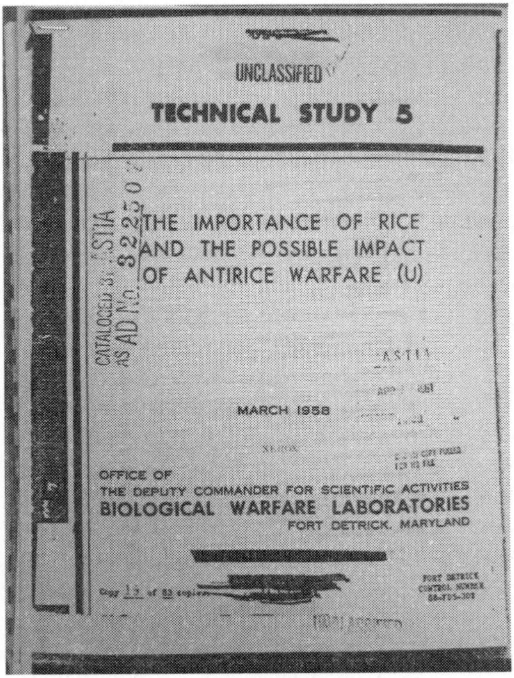
UNCLASSIFIED

TECHNICAL STUDY 5

THE IMPORTANCE OF RICE
AND THE POSSIBLE IMPACT
OF ANTIRICE WARFARE (U)

MARCH 1958

OFFICE OF
THE DEPUTY COMMANDER FOR SCIENTIFIC ACTIVITIES
BIOLOGICAL WARFARE LABORATORIES
FORT DETRICK, MARYLAND

[redacted]

**Figure 5.2** *Front cover of a US study of anti-rice biological warfare. This huge 185-page document reviews rice production, possible anti-rice agents and possible targets for such weapons in China. It analyses, in particular, the way in which the main rice-growing areas might be targeted. The inside front page has a quotation from the Chinese: "men refuse to work, soldiers refuse to fight ... when they are hungry". The study suggests how such conditions might be created through anti-rice biological warfare.*

of an extremely limited range. Such monocultures are particularly vulnerable to attack with biological agents.

## agent production and dissemination

An attacker who has obtained a pathogen with the required characteristics for the purpose intended still faces considerable difficulties – for example, in mass-producing the agent and effectively disseminating it. But, though effective dissemination is difficult if a massive aerosolized attack is intended, production may not be quite so difficult.

In the early 1990s the US Congress Office of Technology Assessment (OTA) did a detailed open analysis of the technologies underlying weapons of mass destruction.[5] For nuclear weapons, they concluded that the "[c]heapest overt production route for one bomb per year, with no international controls, is about $200 million". For chemical weapons, they concluded that an "[a]rsenal for substantial military capability [is] likely to cost tens of millions of dollars". In regard to biological weapons, however, their opinion was that "[e]nough for a large arsenal may cost less than $10 million". There is therefore a great difference in the resources required to obtain nuclear as against biological weapons of mass destruction capability.

To produce something like the anthrax bacterium, there is a need for the seed stock (a small amount of the pathogen) and for standard fermenters such as those used in industry for the production of yoghurt, beer, antibiotics and vaccines (Figure 5.3). According to the OTA, in 1943 a pilot anthrax plant became operational at Fort Detrick, Maryland. It was staffed by five hundred scientists, engineers and technicians. Based on the experience of running this plant, "the decision was made to build a full-scale plant at Vigo, Indiana, at a cost of $8 million, where 1,000 workers would manufacture more than 500,000 anthrax bombs a month". The plant was completed but never actually went into production because the war had ended. However, the scale of productive capabilities was impressive.

Fortunately, all the evidence in the open literature strongly suggests that it is very difficult to achieve effective distribution of an agent in order to cause mass human casualties. According to the OTA, the technical hurdles are as follows:

- The munition or delivery system must generate a cloud of aerosol particles with dimensions that allow them to be inhaled deep into the lungs of the target personnel.
- The agent must be physically stabilized so that it can survive the process of dissemination long enough to infect the target population.
- The agent must disseminate slowly while avoiding loss of viability or toxicity.
- The overall size and shape of the aerosol cloud and the concentration of agent within it must be reasonably predictable so that the dispersion pattern can be matched to the target.

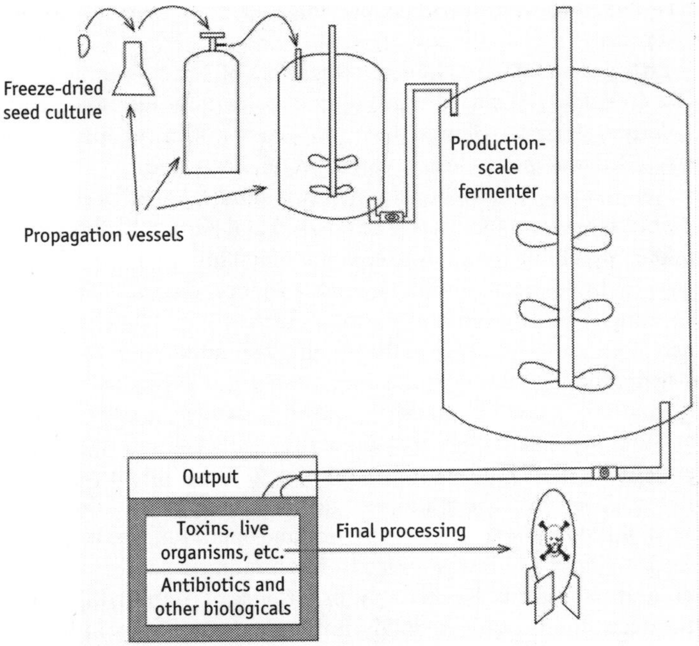


**Figure 5.3** *Diagrammatic representation of production of biological agents by fermentation. This diagram, from OTA, op. cit., shows how a small seed stock of a bacterial agent can be grown in propagation vessels and then in a production-sized fermenter in order to deliver quantities of biological agent. Note that the production of viral agents is generally more difficult and requires the use of more specialized equipment. Source: US Senate, Committee on Governmental Affairs, "Global Spread of Chemical and Biological Weapons", 101st Congress, 1st session, 9 February 1989 (S. Hrg. 101–794), p. 241.*

Though these demanding technical hurdles have been overcome several times in state-level offensive biological weapons programmes, they appear still to be beyond the capabilities of sub-state (terrorist or criminal) groups.

## conclusions

This chapter has given an overview of some of the main potential biological weapons agents that could be used to attack humans, animals and plants in various ways. It has also discussed something of the history of the diseases, the biology of the pathogens and their scientific investigation over the last century and a half. The chapter has also reviewed some of the key thinking behind the choice of the various pathogens that have been, and are, considered to be key threats in the hands of military forces or terrorists.

We now turn our attention to the question of what impact the huge current revolution in the life sciences – the genomics revolution – is likely to have on the possibilities of new biological weapons being developed for offensive purposes and defended against.

## references


1. Rotz, L.D., Khan, A.S., Lillibridge, S.R., Ostroff, S.M. and Hughes, J.M. (2002) Public health assessment of potential biological terrorism agents. *Emerging Infectious Diseases*, **8**(2), 225–30.
2. NIAID (2003) *NIAID Biodefense Research Agenda for Category B and C Priority Pathogens.* Washington.
3. Data on agents have been gathered from a wide range of sources including, in particular: Dando, M.R. (1994) *Biological Warfare in the 21st Century.* London: Brassey's; World Health Organization (2004) *Public Health Response to Biological and Chemical Weapons*, 2nd edn. Geneva; a series of papers on key agents that was published in the *Journal of the American Medical Association* between 1999 and 2002; Kiple, K.F. (ed.) (1993) *The Cambridge World History of Human Disease.* Cambridge, England: Cambridge University Press; Whitby, S. (2002) *Biological Warfare against Crops.* Basingstoke, England: Palgrave.
4. South Africa (1997) *Plant Pathogens Important for the BWC.* BWC/Ad Hoc Group/WP.124. United Nations, 3 March.
5. Office of Technology Assessment (1993) *Technologies Underlying Weapons of Mass Destruction.* Washington: US Congress.

chapter six

# the impact of the biotechnology revolution

## introduction

The use of biological and chemical weapons was banned by the 1925 Geneva Protocol (see Chapter 2). This longstanding agreement, to which most states are now party, has become widely regarded as customary international law binding on all states. As we shall see in Chapter 9, it was supplemented when the Biological and Toxin Weapons Convention came into force in 1975. The BTWC, in its core Article I, further prohibits states from developing, producing, stockpiling or otherwise acquiring or retaining "Microbial or other biological agents, or toxins whatever their origin or method of production, *of types and in quantities that have no justification for prophylactic, protective or other peaceful purposes*" (emphasis added). The emphasized phrase has become known as the "General Purpose Criterion" and it clearly indicates that non-peaceful uses of biological agents are prohibited both now and in the future, with no exception. The later Chemical Weapons Convention (CWC), which entered into force in 1997, has a similar General Purpose Criterion relating to chemical weapons.

When the BTWC was negotiated in the late 1960s and early 1970s the revolution in biotechnology had only just begun. Watson and Crick had only recently, in 1953, discovered the structure of DNA, and the first demonstration that functional genes could be moved between different species (genetic engineering) had only just been conducted.[1] The negotiators of the BTWC could therefore have missed the potential impact of scientific and technological change, but they did not.

Article XII of the BTWC required that a Review Conference be held five years after the entry into force of the convention. Reviews have subsequently been implemented, as the Five-Yearly Review Conferences of 1980, 1986, 1991, 1996 and 2001–2. Article XII of the convention required that "[S]uch review shall take into account any new scientific and technological developments relevant to the Convention." The increasing number of states party to the BTWC – now over 150 – have had the opportunity to assess the ongoing changes in science and technology and to reaffirm that the General Purpose Criterion is not in any way affected by these changes.

The chronological series of reviews and their agreed final statements allow us to follow the growing apprehension among states about relevant scientific and technological developments in the life sciences. This particular aspect of the review conferences will be discussed first here (the broader aspects are dealt with in Chapter 9). Unfortunately, the review system very nearly broke down completely in the extremely difficult circumstances of the 2001–2 Fifth Review Conference, after the United States rejected the verification protocol that had been negotiated over the previous decade. Worse still, the years since have witnessed the publication of a series of experiments that, although carried out for peaceful purposes, indicate major new possibilities for hostile misuse. Some of these experiments and their implications will be discussed. Finally, an attempt will be made to sketch out the probable developments that could occur if we allow an offence/defence arms race in biological warfare to get underway.

## the BTWC review conferences

The BTWC Review Conferences offer two types of documentary evidence on the views of the states that are party to the convention. Firstly, the states are invited to make contributions to a background paper on relevant developments. This paper is then circulated before the review to all states involved. Though most do not choose to contribute, some regularly do so and the contributions often provide much detail on the reasons for their views. The series of background papers on scientific and technological developments are therefore crucial to understanding the thinking of some influential states over the period since the convention entered into force. Secondly, it is usual – though this was not possible in 2001–2 – for an agreed Final Declaration to be produced, summarizing the agreements reached during the Review Conference. The sections of the declaration dealing with scientific and technological changes therefore demonstrate what consensus could be achieved on the evidence available at a particular time.[2] Since the basic General Purpose Criterion is set out in the core Article I, it is in the sections of the declaration relating to Article I where most relevant agreements are reported.

At the *First Review Conference* in 1980 the three depositary states – the United States, the Soviet Union and the United Kingdom – produced a relatively sanguine background paper on the scientific and technological developments and the conference expressed its belief that the provisions of Article I had "proved sufficiently comprehensive to have covered recent scientific and technological developments relevant to the convention". This lack of concern clearly reflected the depositaries' joint background paper, which concluded that, "From a scientific and technological standpoint, the developments discussed in this paper, which are devoted to peaceful purposes, do not appear to alter substantially capabilities or incentives for the development or production of biological or toxin weapons," and that any new agents produced by the novel genetic manipulation techniques would be "unlikely to have advantages over known natural agents sufficient to provide compelling new motives for illegal production or military use in the foreseeable future". That comforting perspective was not to last long as national defence authorities continued to analyse the impact of the new biotechnology (Figure 6.1).

By 1986, the time of the *Second Review Conference*, it was clear that the gene for a toxin might be introduced into a simple bacterium like *Escherichia coli* and then large amounts of toxin produced by industrial-scale fermentation of the genetically altered bacterium. The conference took note of the increasing danger that toxins would be misused, by agreeing to reaffirm in regard to Article I that

> the Convention unequivocally applies to all natural or artificially created microbial or other biological agents or toxins whatever their origin or method of production. *Consequently, toxins (both proteinaceous and non-proteinaceous) of a microbial, animal or vegetable nature and their synthetically produced analogues are covered.* (Emphasis added)

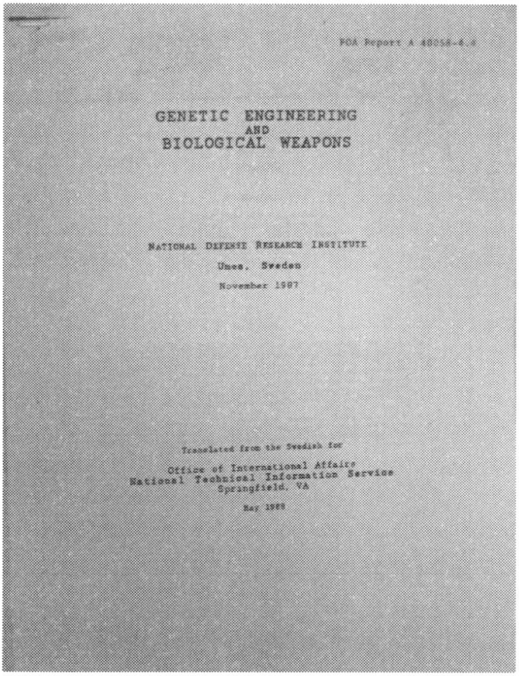

FOA Report A

GENETIC ENGINEERING
AND
BIOLOGICAL WEAPONS

NATIONAL DEFENSE RESEARCH INSTITUTE
Umea, Sweden
November 1987

Translated from the Swedish for

Office of International Affairs
National Technical Information Service
Springfield, VA

May 1988

**Figure 6.1** *Front cover of a Swedish defence study translated into English in the US. This 150-page 1987 study concluded in part that with the development of genetic engineering "the potential threat from more complex, unique, very toxic and hitherto unknown pathogens has become a reality". The perspective of those charged with the defence against biological weapons was rapidly changing.*

At the 1991 *Third Review Conference* note was also taken of the increasing possibilities for genetic manipulation by insertion of the word "altered" into the first sentence so that it read, "the Convention unequivocally covers all microbial agents or toxins, naturally or artificially created or altered, whatever their origin or method of production". The increasing range of concerns at the 1991 review is demonstrated, for example, in the United States' contribution to the background paper on the issue of peptide bioregulators. It pointed out that these can be active at extremely low concentrations and can be modified as agonists (more active than the original) or antagonists (having a contrary action to the original). The paper then stated "Their range of activity covers the entire living system, from mental processes (e.g. endorphins) to many aspects of health such as control of mood, consciousness, temperature control, sleep, or emotions, exerting regulatory effects on the body." Crucially, the contribution continued, "Even a small imbalance in these natural substances could have serious consequences, inducing fear, fatigue, depression or incapacitation." The contribution then adds the warning that "[t]hese substances would be extremely difficult to detect but could cause serious consequences or even death if used improperly".

It is also of interest that the 1991 review agreed a paragraph in relation to Article I which was clearly intended to point out the dangers of field testing of agents. The Final Declaration stated:

> The Conference notes that experimentation involving open-air release of pathogens or toxins harmful to man, animals or plants that has no justification for prophylactic, protective or other peaceful purposes is inconsistent with the undertakings contained in Article I.

Clearly, if an offensive biological weapons programme were undertaken in earnest, open-air field trials with agents would be required to test their effectiveness. The Review Conference presumably had concerns that led to the reiteration of this point following the 1991 Gulf war.

The background paper for the 1996 *Fourth Review Conference* again demonstrated the widening concerns over possible misuse of scientific and technological developments relevant to Article I. The Final Declaration had a total of nine paragraphs related to Article I. Those dealing directly with scientific and technological developments are set out here:

> 5. The Conference also reaffirms that the Convention unequivocally covers all microbial or other biological agents or toxins, naturally or artificially created or altered, as well as their components, whatever their origin or method of production, of types and in quantities that have no justification for prophylactic, protective or other peaceful purposes.
>
> 6. The Conference, conscious of apprehensions arising from relevant scientific and technological developments, inter alia, in the fields of microbiology, biotechnology, molecular biology, genetic

engineering, and any applications resulting from genome studies, and the possibilities of their use for purposes inconsistent with the provisions of this Convention, reaffirms that the undertaking given by the States Parties in Article I applies to all such developments.

7. The Conference notes that experimentation involving open-air release of pathogens or toxins harmful to man, animals or plants that have no justification for prophylactic, protective or other peaceful purposes is inconsistent with the undertakings contained in Article I.

Because a Final Declaration could not be reached at the Fifth Review Conference, this is the most recent statement we have of the consensus views of the states party to the BTWC about the impact of the revolution in the life sciences on the prohibition embodied in the convention.

Even though the *Fifth Review Conference* of 2001–2 was unable to agree a Final Declaration reviewing the whole of the BTWC, it was remarkable nevertheless – in regard to scientific and technological changes – in that the United Kingdom had produced a twenty-nine-page contribution to the prior background paper.[3] The very wide range of topics covered by this contribution emphasized the scope of the changes taking place:

- genomics and proteomics
- bioinformatics
- Human Genome Project and human diversity
- gene therapy
- virulence and pathogenicity
- vaccines and novel therapies
- recombinant protein expression
- toxins and other bioactive molecules
- human infectious disease patterns
- smallpox destruction
- drug resistance
- disease in agriculture
- pest control in agriculture
- molecular biology applications and crops
- trends in protein production technologies
- means of delivery of agents or toxins
- use of pathogens to control weeds and "criminal" crops
- bioremediation: the destruction of material

On the pace of change, the British paper noted that:

> Throughout the various studies and consultations carried out by the UK to inform this review, *it has been clear that the rate of change in science and technology fields relevant to the BTWC has been much greater than in the previous five year period, that is between the third and fourth Review Conferences.* A number of advances in scientific knowledge and its applications could be of consequence for the provisions of the BTWC. (Emphasis added)

Clearly, there were even stronger grounds in 2001–2 for the apprehensions recorded in the Final Declaration of 1996.

## more obvious official concerns

The Five-Year Review Conferences of the BTWC took place among diplomats in Geneva and were little known to the general public and of little interest to the mass media. As the end of the 1990s approached, however, official concerns about biological warfare and biological terrorism became more obvious. The US Secretary of Defense, William Cohen, for example, gave a public warning about the dangers of an anthrax attack.

The US Department of Defense produced a report on the proliferation of biological weapons which specifically considered the dangers of genetic manipulation to make standard agents of more use to attackers.[4] The examples given included the modification of agents to make them resistant to antibiotics or vaccines, and the introduction of the gene for a toxin or bioregulator into another infectious micro-organism. With hindsight it is not difficult to see how growing information about the offensive biological weapons programme in the former Soviet Union during the later stages of the East – West cold war could have led to these concerns (see Chapter 4).

Two examples of work carried out by Soviet scientists at that time, which both found their way into the open literature, are illustrative. The first paper was published under the seemingly innocuous title of "Expression of Cereolysine AB Genes in *Bacillus anthracis* Vaccine Strain Ensures Protection against Experimental Hemolytic Anthrax Infection". It thus appeared as if an improvement in a vaccine had been made. The authors reportedly argued that they were concerned that genes might be transferred between the related

*Bacillus anthracis* and *Bacillus cereus* species in the soil, conferring extra properties on *Bacillus anthracis* which could be countered by the addition of cereolysine AB genes to the anthrax vaccine. However, when a report of this work first appeared at a conference in Winchester, England, it was said to have caused consternation in Washington. Closer inspection of the paper showed that what was of real interest was that golden hamsters vaccinated with the standard *Bacillus anthracis* vaccine used in Russia were *not* protected against a *Bacillus anthracis* strain that had been modified by the addition of the genes from *Bacillus cereus* which enabled this species to break open blood cells. Although the mechanism by which the genetically modified anthrax overcame the vaccine was not elucidated in the paper, the question was raised of whether the standard vaccine used in the West could also be overcome by the addition of such genes.

So the first paper clearly demonstrated that the vaccine for a major biological warfare agent could be rendered useless by genetic engineering of the agent. The second paper went under an even more obscure title, "Additive Synthesis of Regulatory Peptide *in Vivo*: The Introduction of the Vaccine Strain of *Francisella tularensis* Producing Beta-Endorphin". *Francisella tularensis* is the causative agent of tularemia and has long been regarded as a major candidate for a biological weapons agent (see Chapter 5). Beta-endorphin is a well-known bioregulator, and the paper showed that injection of this bioregulator alone into mice produced a short-term reduction in their response to pain, and also muscular rigidity and catatonia. As the bioregulator was quickly broken down by enzymes in the body, these effects wore off within an hour. This paper, however, went on to show that if the gene for beta-endorphin was engineered into the tularemia vaccine strain (the strain used to produce the vaccine) and if that modified strain was then used to infect mice, the effects were stronger and more long-lasting. This could be explained on the basis that the modified tularemia was producing beta-endorphin as it grew and multiplied and therefore that the mice's body enzymes were not able to clear the excess amounts of bioregulator. The bioregulator therefore continued to exert its effects.

If the demonstration of the effective addition of a gene for a bioregulator to a biological warfare agent was not enough, the paper ended by extrapolating from its findings with this statement: "Moreover, it is possible to aim the regulators directly at the target organs by choosing microbes with tropism for the required organs and tissues." If such specific targeting is possible then the amount of the bioregulator required to produce an effect could be much reduced, since enzymatic degradation would not take place during transport of the bioregulator to the organ targeted. That raises the spectre of very small amounts of the modified microbe being required to achieve the malign consequences intended. This approach would be very attractive to a bioweaponeer.

## new-century civil science

After the turn of the century a series of experiments carried out by civil scientists in the West began to raise real concerns in the media, and among some scientists and the general public, that what was being done and published might inadvertently make it very much easier for those with malign intent to achieve their objectives. These worries were obviously greatly heightened by the large-scale disruption and economic costs resulting from the use of a very small amount of anthrax in letters mailed in the United States following the terrorist attacks of September 11, 2001. Some of the well-known controversial experiments, and some that are little known to the general public, will be reviewed before an attempt is made to draw a general picture of the implications.

In January 2001 the London *New Scientist*, which has a wide readership in the scientific community, carried an editorial titled, "The Genie Is out: Biotech Has Just Sprung a Nasty Surprise". This alarming editorial referred directly to the main article in the magazine which, even more alarmingly, stated, "Disaster in the Making: An Engineered Mouse Virus Leaves Us One Step Away from the Ultimate Bioweapon". The article was headed by a picture of a scientist at work and the caption to the picture gave a clue to the concerns raised: "Too close for comfort: the killer virus is a relative of smallpox." The story was obviously important given the history of smallpox and its potential as a bioweapon. Five years previously, *New Scientist* had asked biomedical researchers if they thought it would be possible to make a virus or a bacterium more virulent than the worst natural agents by means of genetic engineering. The reply the magazine generally received was that, though not impossible, it would be very difficult and take a colossal amount of research. As the editorial in January 2001 pointed out, what the replies omitted to mention was the unexpected.

In Australia there is a problem with plagues of mice. Australian researchers had been investigating whether it was possible to prevent such explosions in the mouse population by use of a mousepox virus. They had altered the mousepox genome (hereditary material) by inserting a gene for a mouse egg protein in the hope that this would trigger an antibody response to the egg protein subsequently produced by the virus when it infected mice. This antibody response would then also be raised against the same protein in mice eggs, leading to the mice rejecting their own eggs. When the antibody response was inadequate for the researchers' purposes, they decided to also add the gene for mouse IL-4 (interleukin-4) to the virus. Production of this bioregulator by the modified virus was thought likely to enhance the antibody response and thus the likelihood of egg rejection.

Different strains of mice have different levels of resistance to mousepox. The researchers found that all the experimental animals from a strain that is normally resistant to the virus were killed by the modified virus with the gene for IL-4 inserted. The immune system of the body acts both by producing chemical antibodies and by having a cellular component that is crucial for killing off virus-infected cells. The IL-4 enhanced the chemical antibody production as expected but also shut down the cell-killing component of the immune response. The growth of the virus was therefore unhindered by the immune system and the mice died. Worse still, the researchers discovered that even if this resistant strain of mice were vaccinated beforehand, the modified virus still killed sixty per cent of them.

The researchers recognized the dangers – if someone were to insert human IL-4 into smallpox, the same kind of results might well be obtained, with prior vaccination being unable to prevent people dying from deliberate smallpox infection. In view of the dangers, the researchers consulted their government before publishing their work. However, they eventually decided that it was better to put their results into the open literature and they published a scientific account in a standard journal. The publication sparked a debate in some parts of the security and scientific community, some people stressing the dangers, others asking whether the experimental outcome was really such a surprise and still others arguing that at least we now know that we have to find new means of dealing with viruses besides vaccination.

Though the Australian researchers may have stumbled into the dangers of modified mousepox, the same cannot be said for later work reported in 2003 by American researchers who showed that it was possible to enhance mousepox production of IL-4 by inserting its gene in a better position and adding a promoter sequence for it to the viral genome. The resulting modified virus killed one hundred per cent of vaccinated mice, even when they were treated with the antiviral drug cidofovir. Some infected mice could be saved by the use of a monoclonal antibody that mopped up IL-4. The American scientists were reported to be carrying out similar experiments with cowpox. The original Australian researchers were said to be concerned about the cowpox studies because this virus can attack different animal species including human beings.

The original *New Scientist* editorial had noted that it was possible to hold differing views on whether the work on mousepox should have been done and whether it should have been openly published. What was not possible was for the scientific community to carry on regardless when the rest of society was going to demand that attention be given to the potential dangers as well as the potential benefits of such research work.

As the number of microbial DNA sequences that were fully described increased in the later 1990s, some of the scientists involved began to speculate as to whether an organism might be constructed synthetically. Some believed that this could still be quite difficult to do. However, just eighteen months after the mousepox work was reported, American researchers announced that they had synthesized the simple poliovirus. The poliovirus has a short RNA genome and replicates in a very simple way when it gets inside cells. The sequence of the genome is known and is available on the internet. Bits of DNA complementary to viral RNA sequences were bought from ordinary companies that provide such material for commercial reasons and these pieces were joined together by standard laboratory methods. An RNA polymerase enzyme was then used to transcribe the complementary DNA into viral RNA, which was then placed in a cell-free extract where it replicated into infectious poliovirus.

Again this research and its publication sparked a debate within the scientific community. Some eminent scientists such as Craig Venter, the US scientist who played a leading role in the Human Genome Project, said it was irresponsible to synthesize a human pathogen and without any scientific justification. Others argued that it had been known for two decades that such a synthesis was theoretically possible and the practical demonstration of this fact added nothing new. The synthesis of polio was carried out over a period of many months. However, the scientific paper describing the polio synthesis ended by noting ominously that, given ongoing scientific advances, much more rapid syntheses of such viruses would be possible in the future. Great efforts are being made internationally to finally eliminate the disease of polio and, though polio is not regarded as a major potential bioweapons threat, it is thought possible by some researchers that the synthesis could indicate how to make more complex viruses such as Ebola which could pose a distinct bioweapons threat.

Ironically, it was Venter's laboratory that showed just how easily the original techniques could be developed. By using better methods, for example to filter out erroneous stretches of the DNA they had purchased, Venter's group managed to synthesize a simple bacteriophage (a virus that attacks bacteria) within a couple of weeks rather than the many months it took for the poliovirus synthesis. Venter's publication of the work on the bacteriophage stated that his group had had an independent ethical review carried out before they did the work and that there was clearly a lag between genome synthesis methods and genome sequencing methods. The paper ended by looking to a future in which that gap was closed: "Synthetic genomics will become commonplace and will provide the potential for a vast array of new and complex chemistries altering our approaches to production of energy, pharmaceuticals, and textiles." Again there was debate about whether this was sensible scientific work, perhaps leading to the artificial creation of microbes that could be used to clean up waste, or whether the work could be misused to create bioweapons more easily.

So far, our examples have shown how a pathogenic micro-organism could be made much more lethal and how a pathogenic micro-organism could be synthetically created. In the mousepox example, we briefly discussed the body's main defence mechanism – the immune system. As we now understand from studies on AIDS, if the immune system is successfully attacked the balance of advantage swings very much towards the attacking pathogen. A third experiment, also published in 2002 about eighteen months after the mousepox experiment first surfaced, more directly raised concerns about means of attacking the immune system. Most worryingly, this example concerned an investigation of how smallpox disrupts an early stage in the immune system's response. Early on in an invasion by a pathogen like smallpox the body's system of proteins termed "complement" attack the virus. Important components of this complement system are called C3b and C4b.

In order to vaccinate people against the dangerous smallpox virus (variola) a closely related but not dangerous virus called "vaccinia" is used. This induces immunity against smallpox and in people with competent immune systems does not cause any disease. Poxviruses like smallpox have evolved ways and means of evading the actions of the immune system, and the investigators in the third experiment showed that the vaccinia vaccine strain produces a protein that inactivates the C3b and C4b complement proteins. They called this protein "vaccinia virus complement protein" (VCP). As smallpox virus was not available to the researchers, they scanned the genome databases of various strains of the deadly variola (smallpox) virus and discovered that they all contained coding for a similar protein to the vaccinia VCP. They called this homologous smallpox protein "smallpox inhibitor of complement system" (SPICE).

Critically, their further investigations showed that "SPICE is nearly 100-fold more potent than VCP at inactivating human C3b and 6-fold more potent at inactivating C4b." They commented that the lesser effectiveness of VCP compared with SPICE could be one reason why vaccinia causes a much lower mortality. So the work indicates one reason why variola is much more lethal than vaccinia but raises the question of whether such knowledge might be misused by those with malign intent to create an even more lethal smallpox virus in the future.

Recreation of a deadly virus is not as farfetched a possibility as we might have imagined just a few years ago. The Spanish flu epidemic of 1918–19 is thought to have killed at least twenty million people worldwide and it is considered to be the worst pandemic ever to have affected the human species. Furthermore, whereas influenza is usually more serious for the very young and for older people, this pandemic was unusual in drastically affecting many young adults.

In the late 1990s investigators began to use modern molecular biology techniques successfully to recover and identify parts of the genome of Spanish flu. They worked particularly with tissues that had been retained from post-mortem examinations of people who had died of the infection. The fixation methods used on the tissues at the time of the epidemic were crude, so this was a difficult technical operation. Gradually, however, a series of papers was published which showed the particular sequence of crucial genes of this organism. Influenza virus has an RNA genome and an initial paper in 1997 reported, "RNA from a victim of the 1918 pandemic was isolated from a formalin-fixed, paraffin-embedded, lung tissue sample. Nine fragments of viral RNA were sequenced." By 2004, however, a paper in the major journal *Nature* was reporting that the complete sequencing of several genes of the 1918 virus had made it possible to study the functions of the proteins coded for by these genes and to identify some of the properties of the virus which might have caused its extreme virulence. In particular, the paper stated,[5] "Here we demonstrate that the HA [haemagglutinin] of the 1918 virus confers enhanced pathogenicity in mice to recent human viruses that are otherwise non-pathogenic in this host." So, adding the particular gene for the 1918 type of HA protein to recent influenza viruses, which normally do not affect mice badly, makes these viruses more pathogenic in mice. It is known that the influenza virus genome is unstable and that we encounter slightly different variants in most years as well as occasionally those with major changes which produce pandemics. This means that there are many known variants for investigators to manipulate. What was striking about the modified viruses with the HA from the 1918 influenza was that they were able to severely attack the lungs of mice in precisely the manner that was the hallmark of the 1918 illness in humans.

The investigators involved in this work argue that they are using modern techniques prudently in order that we may better understand the 1918 pandemic, which may, in turn, enable us to develop countermeasures against this strain should it reappear. Others have argued that there are plenty of other strains of influenza for study and that this work could create a roadmap for those who might wish to recreate the 1918 virus for malign purposes. Some people may feel that to consider the genetic manipulation of pathogens to make them more virulent or the creation of new – or recreation of old – pathogens is not a likely route for bioweaponeers to take. After all, one may well argue, anthrax, Ebola virus, plague and so on are surely bad enough. Why should anyone seek out new pathogens when there are so many readily available? Nevertheless, some in the military are already worrying about such possible future threats.

A paper entitled "Genome Projects and Gene Therapy: Gateway to Next Generation Biological Weapons" appeared in Volume 168 of the scientific journal *Military Medicine* in November 2003 and noted:[6]

> Genes that are essential to life are being identified as are DNA sequences that are responsible for regulating these genes.

> Disruption of these genes may be impairing so they may be lucrative targets for *genomic warfare*. In addition, genes from other organisms which can produce products responsible for illnesses in man are being identified and these may also be weaponised. (Emphasis added)

The author here is using the term "genomic warfare" because he believes that all parts of the genome – protein-coding regions and control sequences – could be targets for future warfare. The paper gives much space to advances in gene therapy. These techniques for manipulating the genome are being developed for sound medical reasons but, as the author points out, they could also be misused in future genomic warfare.

## defence/offence interaction

The currently increasing investments in biodefence are mainly designed to counter the use of the well-known biological agents that were often weaponized in the past. As there are only a limited number of such agents, the defence is likely to be able eventually to achieve its aims. However, should an arms race take place, offence would then be likely, in order to overcome such defence, to develop the kinds of modified agents we have discussed in this chapter. Only so many such manipulations are possible, so we can expect that, in time, the defence will again be able to gain the upper hand.

The problems that the defence would subsequently face could then become impossible to handle.[7] As the century progresses, more and more of life's fundamental processes will become open to manipulation for good – and malevolent – reasons. So the attackers could switch from concentration on the agent to concentration on the target. Given the huge range of potential targets in the physiological systems of living organisms, and the many ways in which each might be attacked, the advantage will shift again and remain for a long time with the attacker. Even those who seek generic means of protecting the body against multiple kinds of attack by pathogens have to accept that it is a simpler task to do harm than to provide sure and safe protection.

Finally, it would be wrong to think that only pathogens and genomics could be subject to malign manipulation in the future. Major advances are being made in many areas of the life and associated sciences and technologies. George Poste, who has tried to alert his fellow scientists to the dangers in recent years, has pointed to the "brain bomb". As our understanding of the nervous system continues to develop, he suggests, we will be able to carry out all manner of malign manipulations as well as to use the knowledge to help those suffering from mental illness.[8]

## conclusions

The life sciences are in a period of very rapid development. This new knowledge may help us to prevent and deal with biological warfare and terrorism by giving us better means of detection, protection and treatment. Yet it could also be misused, and this is not knowledge that – as in the case of nuclear weapons – only a few with huge resources can misuse. It is clear that the technology is becoming simpler and spreading wider. Preventing the malign misuse of this burgeoning technology will be a major task in coming decades.

## references


1. Guttman, B. (2002) *Genetics: A Beginner's Guide.* Oxford: Oneworld.
2. Dando, M.R. (1994) *Biological Warfare in the 21st Century: Biotechnology and the Proliferation of Biological Weapons.* London: Brassey's.
3. Nathanson, V. and Dando, M.R. (2004) *Biotechnology, Weapons and Humanity II.* London: British Medical Association.
4. See reference 3 for details of the scientific experiments discussed.
5. Kobasa, D. et al. (2004) Enhanced virulence of influenza A virus with the haemagglutinin of the 1918 pandemic virus. *Nature*, **431**, 703–7.
6. Black III, Col. J.L. (2003) Genome projects and gene therapy: gateways to next generation biological weapons. *Military Medicine*, **168**, 864–8.
7. Petro, J.B., Plasse, T.R. and McNulty, J.A. (2003) Biotechnology: impact on biological warfare and biodefense. *Biosecurity and Bioterrorism*, **1**(3), 161–8.
8. Poste, G. (2002) Advances in biotechnology: promise or peril. <www.hopkins-defense.org/sympcast/transcripts/trans_poste.html>.

chapter seven

# attack scenarios today

## introduction

In this chapter we consider what is known of the consequences of attacks using biological weapons.[1] We cover both anti-personnel and anti-agriculture attacks and pay particular attention to recent discussions of potential attacks by terrorist groups using biological weapons. For such terrorist attacks, the probable difficulties of effectively aerosolizing the biological agents described in Chapter 5 should be kept in mind.

## anti-personnel attacks

First to be considered is the well-known literature on the use of biological weapons as weapons of mass destruction (WMD) by spreading an agent in an effective manner through the air over a large area. We will then turn to recent discussions of potential terrorist attacks on both a large and medium scale. It should be borne in mind that biological agents can be used for assassination. Georgi Markov was killed in London by a ricin-laced pellet injected into his leg, and much of the South African offensive chemical and biological weapons programme was directed at developing means of assassination for use against the apartheid regime's enemies (see Chapter 4).

## WMD attacks

In 1969, in the run-up to the eventual negotiation of the Biological and Toxin Weapons Convention, a report on *Chemical and Bacteriological (Biological) Weapons and the Effects of Their Possible Use* was produced by the United Nations Secretary General.[2] In its second chapter the report analysed the probable effects of biological weapons compared with nuclear and chemical weapons. The relevant table is reproduced in part here (Table 7.1). It is immediately obvious from this table that in the right conditions a single bomber could affect a huge area with a biological weapons agent. The area would be much larger than that affected even by a one-megaton nuclear weapon (100,000 km$^2$ as against 300 km$^2$) and the expected death rate would be twenty-five per cent of the victims of the attack.

The report was appropriately guarded in stressing that these probable effects were estimates and much would depend on the prevailing conditions (e.g. the weather) during a biological weapons

**Table 7.1** Probable effects of the use of nuclear, chemical and biological weapons (carried on a single bomber) on an unprotected population*

| | Type of Weapon | | |
|---|---|---|---|
| Criterion | Nuclear *(one megaton)* | Chemical *(15 tons of nerve agent)* | Biological *(10 tons of biological agent)* |
| Area affected | Up to 300 km$^2$ | Up to 60 km$^2$ | Up to 100,000 km$^2$ |
| Time to onset of effect | Seconds | Minutes | Days |
| Damage to structures | Destruction over 100 km$^2$ | None | None |
| Maximum effect on humans | 90% deaths | 50% deaths | 50% morbidity: 25% deaths if no medical intervention |

* From Secretary General, op. cit.

attack, but the people involved in producing the report were of the stature of, for example, Sir Solly Zuckerman, Chief Scientific Adviser to the government of the United Kingdom, who had access to the results of all experimentation done in the British offensive biological weapons programme during and following the Second World War. The report should therefore have left no doubt that biological weapons could be used as weapons of mass destruction.

By the early 1990s it was possible to add to this United Nations report a variety of estimates from other open sources.[3] These all led to the same conclusion – that biological weapons could, if used in an appropriate manner, cause huge levels of disease and death. Consequently, they had to be regarded as weapons of mass destruction equivalent in many ways to nuclear weapons. Some of these estimates are set out in Table 7.2. The first case in the table was described by the Stockholm International Peace Research Institute (SIPRI) in Volume 2 of its classic 1970s study, *The Problem of Chemical and Biological Warfare*. The second case was analysed in great detail by Steve Fetter in the journal *International Security* in 1991 and the third and fourth cases were taken from the 1993 US Office of Technology Assessment study, *Proliferation of Weapons of Mass Destruction: Assessing the Risks*.

What is clear from the first three cases is that they support the view put forward in the 1969 United Nations report, that a biological agent delivered effectively by bomb or missile would cause huge levels of casualties. Furthermore, during the state-level offensive biological weapons programmes of the last century the bioweaponeers made great advances in the effectiveness of their abominable weapons. This is evident if we consider the fourth case in Table 7.2.

It had been shown that the most effective way to use a biological weapon was to spray it in a line (say from a plane) so that the material drifted across the target. This was best done at night so that the ultraviolet (UV) light from the sun did not kill the (anthrax) bacteria. The OTA example assumed that the attack was on Washington, DC. The three scenarios under which the attack was carried out assumed the following weather conditions: a clear, sunny day with a light breeze; an overcast day or night with a moderate wind; and, worst of all, a clear calm night. The consequences are shown as diagrams in Figure 7.1. Analysis of where, and in what quantities, the one hundred kilograms of anthrax spores sprayed in a line on the windward side of the city would land, and the concentrations

**Table 7.2** Some publicly available information about large-scale attacks using biological weapons*

| Study (case) | Weapon system | Area affected ($km^2$) | Fatalities |
|---|---|---|---|
| *I. SIPRI* | | | |
| Bomber | 10 kt nuclear | 30 | n.a. |
| | biological agent | 0–50 | n.a. |
| | VX nerve gas | 0.75 | n.a. |
| | 5–6 t high explosive | 0.22 | n.a. |
| *II. International Security* | | | |
| Missile on sparsely populated city | 20 kt nuclear | n.a. | 40,000 dead and 40,000 injured |
| | 300 kg sarin | n.a. | 200–3,000 |
| | 30 kg anthrax spores | n.a. | 20,000–80,000 |
| *III. OTA* | | | |
| Missile on sparse to moderately populated city | 12.5 kt nuclear | 7.8 | 23,000–80,000 |
| | 300 kg sarin | 0.22 | 20–200 |
| | 30 kg anthrax spores | 10 | 30,000–100,000 |
| *IV. OTA* | | | |
| Line attack | 100 kg anthrax spores | 46 (clear day) | 130,000–460,000 |
| | | 140 (overcast) | 420,000 to 1.4 million |
| | | 360 (clear night) | 1–3 million |

* From Dando, op. cit.

required to infect and kill, suggested that in the worst case some one to three million people would die.

As we saw in Chapter 5, it would be possible to treat people successfully with antibiotics if treatment started early enough. However, whether any public health system could cope with such numbers of casualties seems unlikely, even if the attack was detected. Additionally, if a genetically manipulated organism with built-in antibiotic resistance had been used, even early treatment might not

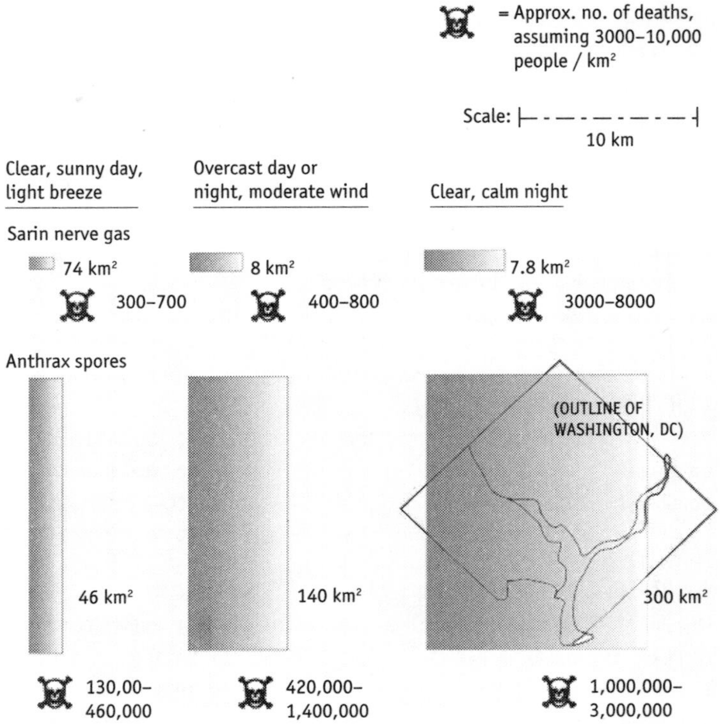


**Figure 7.1** *Comparative lethal areas of chemical and biological weapons. This diagram shows the lethal areas produced by chemical and biological agents dispersed from a highly efficient line source under various meteorological conditions. Note that the respective quantities in each case are one thousand kilograms of sarin nerve gas and one hundred kilograms of anthrax spores. The caption to the original noted that "[m]ore anthrax would be inefficient", highlighting the greater danger posed by anthrax.*

*Source: OTA (1993) Proliferation of Weapons of Mass Destruction: Assessing the Risks. Washington.*

work. We have to accept that biological weapons could be used as weapons of mass destruction even if the actual outcome of an attack might vary considerably in different weather conditions. It will also be noted from Tables 7.1 and 7.2 that the calculated effects of the use of a biological agent like anthrax far exceeds that of a nerve gas. These estimates have been confirmed in many national studies, for example in the United States.

## non-WMD attacks

Also in the run-up to the negotiation of the Biological and Toxin Weapons Convention the World Health Organization (WHO) produced, in 1970, the first edition of its report on *Health Aspects of Chemical and Biological Weapons.*[4] This is of interest because it analysed cases of WMD usage and also smaller-scale attacks using biological weapons.

The WHO report considered WMD-style attacks using tularemia as the biological agent against a city of five million people in a developed country (with good medical resources), on a similar-sized city in a developing country and against a city of 500,000 people, again in both a developed and a developing country. These cases do not change the overall conclusion derived from the earlier case studies discussed here, but they do give more details on the problems that would be encountered by the health and municipal authorities. A similar analysis was made for an attack using plague against a city of five million in a developed country (with antibiotics available) and against a similar-sized city in a developing country (with only a small supply of antibiotics). In the latter case it was suggested that a total death toll of 250,000 could result from the successive waves of epidemic caused by the agent.

The WHO report then went on to consider sabotage of water supplies with biological agents. The authors considered contamination at the intake or treatment works, at the raw or treated water reservoirs and at a transmission main. They dismissed the idea of a takeover (or staff corruption) at the treatment plant, since this would require the attacker to have very special access and favourable conditions – although it would obviously lead to a worst-case scenario. They also dismissed the idea of contamination of a water reservoir as a means of effective sabotage. In their opinion, controlled injection of an agent into a trunk main "would be potentially the most devastating in effect, difficult to prevent and detect, and feasible in practice in many water systems".

Two of the hypothetical cases considered in the report used as the contaminating agent:

(a) the typhoid bacillus, which produces no recognizable symptoms for about 1 week; and
(b) botulinal toxin, type A, which would produce no recognizable symptoms until 6 or 8 hours after ingestion and, in a stabilized form, would resist denaturation by the elements found in a normal water supply.

The report noted that these agents would likely escape detection and could produce effects at low concentration so that only a small amount of initial contaminant would be required. It also suggested that if a chlorine residue were left in the mains, its effect on the agents could be reduced by the simultaneous injection of a dechlorinating agent.

Four particular cases were analysed. The typhoid attack had two sub-cases: an attack on a city of one million plus in a developed country, and the same in a developing country. Likewise the botulinum attack had two sub-cases: an attack on a city of fifty thousand in a developed country, and the same in a developing country. Water consumption in the four sub-cases would be affected by the climate in the different countries as also would be the level of medical care available. So overall there would be a good spread of hypothetical cases in which to examine the consequences of such attacks.

The hypothetical attacks were deemed to take place without warning so that no special precautions were undertaken by the authorities and the attackers were assumed to have knowledge of the mains systems layout so that they could achieve maximum effect. So that impossibly large amounts of agents were not required, it was assumed that one kilogram of freeze-dried culture of typhoid bacillus was used to attack a city with a population of one million plus and for botulinum toxin an attack with 0.24 kilograms of toxin on a city of fifty thousand was considered. If the typhoid bacillus were used, because of the one-week incubation period the authorities could take no immediate remedial action. Following the attack with botulinum toxin, as symptoms began to appear after 6–8 hours, the authorities might be able to tell people not to drink the water until it had been flushed clear of contamination. Two types of consumption were also considered. In the first, for an industrial community in a temperate climate, it was assumed that fifteen per cent of the people would not drink water direct from the tap in the relevant period. In the second, for a city in a tropical area with little industry, it was assumed that everyone would drink the tap water in a 3–4-day period.

The percentages of people who would be infected by different levels of typhoid bacillus is known from other studies of human volunteers. In the first sub-case, of a large industrial city in a temperate climate, it was assumed that the amount of water drunk per person was 0.5 litre per day. The total number of typhoid cases was then calculated to be about thirty-five thousand; assuming that early and effective use of antibiotics reduced the fatality rate to 0.6%, a total number of two hundred deaths was anticipated. In the second sub-case, of the non-industrialized city in a hot climate, the water consumption was assumed to be two litres per person each day. It was calculated that 100,000 micro-organisms would be delivered to each of over 125,000 people (out of the million in the city) and cause many of them to become ill. If no facilities were available for mass treatment, some 4500 people might die.

In the botulinum toxin sub-cases, because of the rapid onset of symptoms and the extreme toxicity of the agent, the number of deaths would be much higher – around thirty thousand people. There is little difference in the death toll between the two botulinum sub-cases because there would be little chance in either the developed or developing country – among the people who receive lethal doses – to apply the modern medical treatments that could be available in a developed country.

These hypothetical examples reinforce conclusions drawn from studies of the consequences when pathogenic contamination occasionally causes problems even in developed countries, which suggest that sabotage with pathogenic micro-organisms or their toxic products cannot be ignored as a threat to the general population. Contamination of the food supply chain or soft drink production, for example, might also be considered by terrorists.

## anti-agriculture attacks

Huge losses in agricultural production are caused by plant pathogens, so it is hardly surprising that the bioweaponeers of the last century gave careful consideration to the deliberate destruction of the staple food crops of their potential enemies. For example, a document released under the United States Freedom of Information Act, dated March 1958, was titled *The Importance of Rice and the Possible Impact of Antirice Warfare* (see Figure 5.2). The 185-page document was a study of how China might be attacked through destruction of its rice crop – thereby reducing the country's "capability and will to wage war". The conclusions to the report noted, among other things, that "A susceptible period for attacking rice exists shortly after transplanting, an operation readily apparent by aerial observation and suitable as a reference for timing an attack." The idea here was that by spreading a biological agent on the crop at this early stage the pathogen would cause a primary infection to grow and disperse spores which would then cause secondary infections. The amount of material the attacker would have to use would therefore be quite small.

The document reported studies on the susceptibility of different rice varieties to different isolates of the chosen fungus. The aim here was to use a mixture of different types of the fungal agent *Piricularia oryzae* so that any variation in the type of rice being grown would not reduce the consequences of an attack. *Piricularia oryzae* causes rice blast; its spores could be easily produced by methods perfected in the United States. The dried spores were stable in storage and resistant to environmental degradation. It is difficult to overstate the scope of the study. The main rice-growing areas of China at the time were analysed, as were the average contributions of the rice crop to the daily calorific intake of the Chinese population. The report suggested that a two-pronged attack would be effective, a biologically active chemical agent being used where the environment was not suitable for use of the pathogen, but "[w]here the environment is suitable for the use of an antirice pathogen it should be used".

The possibility of large-scale attacks on staple crops such as rice therefore has to be taken seriously. As is clear from the discussion of available agents in Chapter 5, it would also be possible to direct attacks at economically important cash crops and at animal husbandry.

## current terrorism concerns

Following the attack on the twin towers in New York and other targets in the United States on September 11, 2001, and the anthrax letter attacks soon after, concerns about bioterrorism grew. An article in the 2002 *Annual Review of Microbiology* encapsulated many people's concerns. Titled "Bioterrorism: From Threat to Reality", the article began:

> The fears and predictions of attacks with biological weapons, which were increasing at the close of the twentieth century, were transformed into reality not long after September 11, 2001, when several anthrax-laden letters were sent through the U.S. postal system.

The article went on to point out that it was fortunate that the material was not used in a massive aerosol release that could have affected

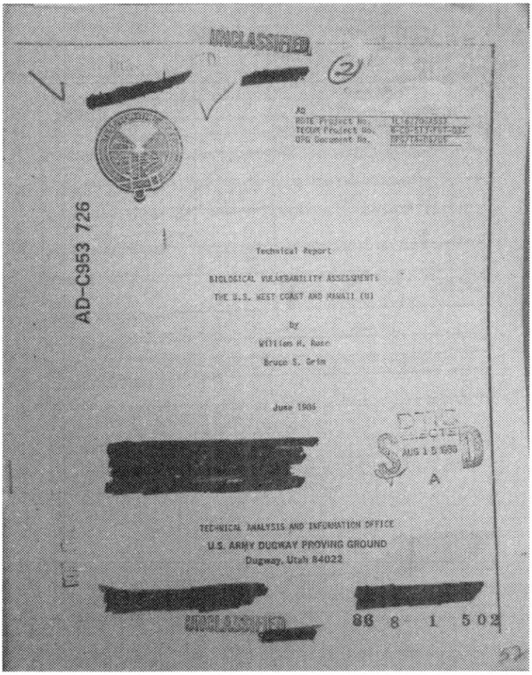
UNCLASSIFIED

[redacted]

AD-C953 726

Technical Report

BIOLOGICAL VULNERABILITY ASSESSMENT:
THE U.S. WEST COAST AND HAWAII (U)

by

[illegible]

[illegible]

June [illegible]

[redacted]

TECHNICAL ANALYSIS AND INFORMATION OFFICE
U.S. ARMY DUGWAY PROVING GROUND
Dugway, Utah 84022

[redacted]

86 8 1 502

**Figure 7.2** *Front cover of a US study of the vulnerability of the US west coast and Hawaii to biological attack. This study considered a variety of scenarios for large-scale biological weapons attacks on the western United States. The vulnerability of the areas and the possible effects of such attacks on the military and civilian populations were analysed.*

thousands, but it also noted that, though the problem was dealt with by the authorities, "[f]ear gripped the nation". Many people were undoubtedly very worried about what might happen if terrorists were to use biological weapons in earnest, and both national and international authorities reacted by reviewing and improving their safety precautions.

It is known that national biodefence communities had long considered the possibility of bioterrorism. In the United States tests had been carried out with pathogen simulants by the US Navy, which released an aerosol off San Francisco and tested how far organisms were carried inland and in what concentrations; simulants had also been released on the New York subway and estimates made of how many people would be infected in a real attack and so on (Figure 7.2). Similar analyses had been carried out in the UK, and, no doubt, in other countries.

A Technical Note[5] written by a member of the US Army at its main biodefence testing site (Dugway Proving Ground) recorded that "The use of biological materials in conjunction with terrorist events has been addressed in various studies and reports (see Annex A)." Annex A to this 1986 report then listed nineteen US Army reports on the subject over the previous eleven years, and a further eight studies by institutes such as the Rand Corporation over the same period. The introduction to the annex noted that the list of reports in the annex was intended to be a representative sample and it pointed out that:

> When, in 1973, DPG [Dugway Proving Ground] received a mission assignment ... to maintain the program for technical assessments of foreign biological threat, one of the first studies was an assessment of the potential threat from the use of biological materials by terrorists. Since that time, consideration of terrorist employment of biologicals ... has been a continuing part of the program.

So in no sense did governments like those of the US and the UK lack information on what terrorists might be able to do in biological attacks.

The second report in Annex A was titled *Covert Biological Weapons Literature Review*.[6] This 1975 report sought to review the possibilities for a subversive group carrying out a biological weapons attack in the United States. It specifically excluded large-scale, direct, aerosolized biological weapons as a means of covert attack by such a group, but concluded:

> BW agents can be used against man, animals, or plants. Reports reviewed here show plainly that a ... subversive group could produce a variety of effective BW agents, and deliver them against the civilian population and agricultural and water resources of the United States by many covert and overt means.

So there can be little doubt that a terrorist group at the present time could carry out a small- to medium-scale biological weapons attack. A massive WMD aerosolized agent attack is quite another matter. All of the technical literature and opinion holds the view that, though the problems of production and dissemination have been solved in state programmes in the past, it is presently still unlikely that a sub-state group would have the necessary capabilities and resources.

Milton Leitenberg of the Center for International and Security Studies in the School of Public Affairs at the University of Maryland has emphasized that[7] "threat assessment, most particularly regarding 'BW terrorism' – the potential for BW use by non-state actors – has been greatly exaggerated". In a paper presented at a meeting organized under the auspices of the Italian Ministry for Foreign Affairs in April 2002 Leitenberg went on to argue that this exaggeration was counter-productive both in suggesting to those with malign intent that they should be interested in biological warfare, and in distorting priorities for investment in public health. So it is necessary to be careful and to keep the threats in realistic proportion as we examine some recent analyses.

## anti-agriculture bioterrorism

As many experts have cogently argued, it is agriculture that is particularly vulnerable at present. Potential attacks on agriculture have been characterized as "[l]ow-tech, high consequence bioterrorism". The reasons for this characterization are not difficult to understand. Many studies suggest that such bioterrorism would require relatively little in the way of specialist knowledge or technical expertise and technology, the diseases are highly contagious to the intended targets (but not to humans) and would therefore spread rapidly and it would cost a great deal of money to eradicate such diseases. An added cost would come from the losses in international trade that would follow as other countries tried to protect themselves from the diseases.

Taking the United States as an example, one recent analysis argued:[8]

> Pathogens that cause diseases such as FMD, rinderpest, African Swine Fever (ASF), soybean rust, Philippine downy mildew of maize, potato wart, and citrus greening, could, if introduced into the continental US, have serious consequences for the US economy.

Reporting that after the first case of FMD was found in the UK outbreak in 2001 the European Union and others immediately blocked imports of British beef, sheep and pigs and the products derived from them, the authors noted that the scale of the US industry is much greater than that of the UK. They concluded that "with $37 billion of beef, $23 billion of dairy and $9.2 billion of pork sales annually … the trade consequences of an outbreak of FMD would be much larger". They pointed out that a recent study using conservative estimates of the impact of an FMD outbreak on Californian agriculture suggested a US $6–13 billion loss – even if confined to just California and eradicated within 5–12 weeks.

Taking another example, they pointed out that even though kernel bunt of wheat, caused by the fungus *Tilletia indica*, does not have a large direct effect on crop yield, about eighty countries ban wheat imports from regions infected with the fungus. The disease was discovered in Arizona in 1996, probably through an accidental introduction from Mexico, and produced an immediate threat to the US $6 billion per year US wheat crop, about fifty per cent of which is exported. The US Department of Agriculture's Animal and Plant Health Service therefore spent some US $60 million to eradicate the disease between 1996 and 1998 and the growers in the small area affected were estimated to have lost over US $100 million in sales and through extra production costs incurred.

Against that kind of background it is obvious that potential attacks against agriculture using biological weapons must be taken seriously. Whether they really are taken sufficiently seriously today is an open question.

## non-WMD attacks on people

A May 2004 report by the US Congressional Research Service was titled *Small-Scale Terrorist Attacks Using Chemical and Biological Agents: An Assessment Framework and Preliminary Comparisons.*[9] The report cautioned against thinking predominantly about the requirements for a state-level offensive programme designed to gain a capability for launching massive aerosolized attacks and achieving WMD-level casualties. Because the information we have in the public record comes from such state-level offensive programmes, it is entirely understandable if our thinking is dominated by state-level requirements. However, the report points out that

> for terrorist distribution of a C/B [chemical/biological] agent, many steps considered to have *high practical difficulties* may be nonexistent in the case of terrorist groups that wish to launch only a small-scale attack and that have a low regard for their personal safety.

It then shows a flow chart for a state programme and a terrorist programme (reproduced here as Figure 7.3). This shows the sequence of steps required for these different types of programme to achieve their objectives.

The report points out that the steps italicized in the state-level programme flow chart would not be needed in the terrorists' programme. The terrorists would not need agents with a long (storage) shelf life; they would not need to optimize the functions of a large-scale dissemination device, to develop rigorous prophylaxis methods or to optimize the manufacture of large amounts of agent. The technical difficulties for the terrorists would clearly be very much reduced. If a terrorist group analysed the major disruptive effects of the very small-scale use of anthrax (albeit lethal) in mail attacks in the US in 2001, it might well decide that a relatively small attack using rather crude technology and at some risk to its own personnel would help to gain its objectives. So small-scale, low-level technology attacks cannot be dismissed as totally unlikely in the future.

The Congressional Research Service report went on to analyse the agents that appear in standard lists such as those of the CDC, but not from the usual perspective of which would be most dangerous coming from a state-level military programme, but rather from a terrorist's perspective. A virus like Marburg, for example, would standardly be viewed as a high-level threat (in Category A; see Chapter 5), but, since it is not easily obtainable in nature, it gets downranked in this perspective. The report notes:

> C/B agents that were considered high threats in other frameworks appear to present a lesser threat when viewed in the small scale attack context. Conversely, C/B agents that were considered of lesser threat when considering mass casualty attacks may be ranked more highly in the small scale context, as barriers to mass use may be missing when the agent is used on a small scale.

Not surprisingly, therefore, the report went on to say, "[b]ecause of the differences, policies designed to protect against catastrophic C/B attack may not provide equivalent protection against small scale C/B attack". Again it is not certain whether this view is widely held and is informing policies designed to counter terrorism.

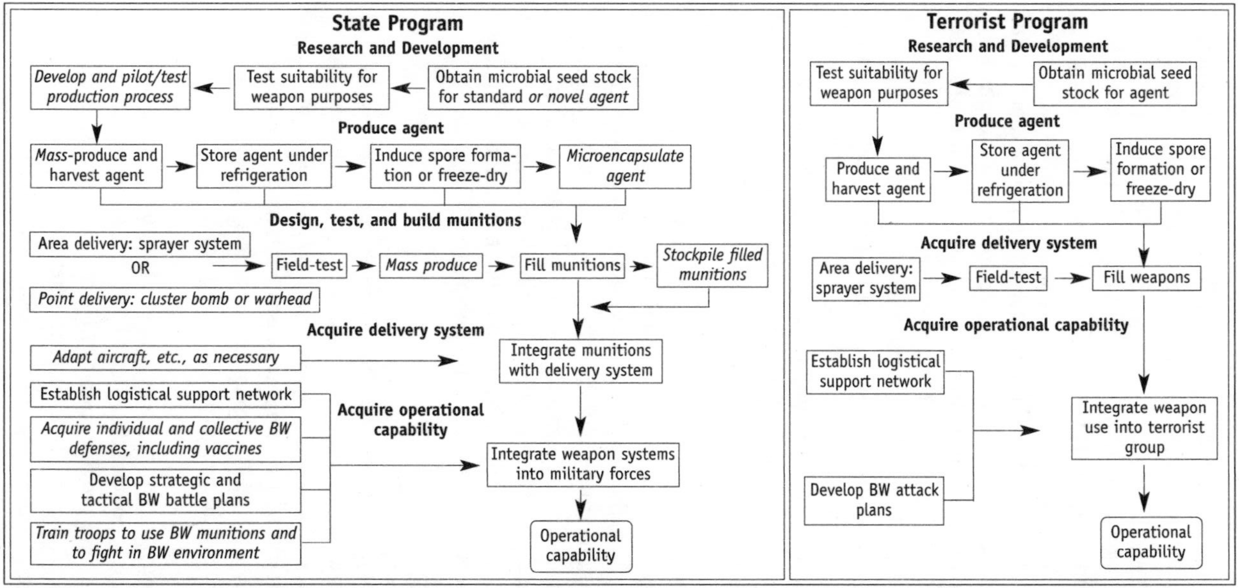


**Figure 7.3** *Comparison of state and terrorist biological weapons development. This flow diagram illustrates the key differences between a state-level offensive biological weapons programme (on the left) intended to produce a WMD capability and that of a terrorist group with lesser ambitions (on the right). The additional capabilities needed in the WMD programme are in italic script on the left and are removed from the right-hand illustration of a terrorist programme.*

## catastrophic bioterrorism

Despite the arguments presented so far in this chapter to the effect that we may not have paid enough attention to biological weapons attacks on agriculture and to smaller-scale attacks on humans, it is obvious from the arguments about offence/defence interaction presented in the last chapter, and from the rapid evolution of biotechnology in general, that we neglect at our peril to think about large-scale attacks in the future. One person who has tried to think this issue through over a period of years is Richard Danzig, the former US Navy Secretary. He has been concerned about what he calls "catastrophic bioterrorism" and he is far from sanguine about the problem.

Danzig believes that the aim of the terrorist is to disable good governance, enhance divisiveness and undermine the confidence of citizens in their government. Writing in August 2003, he argued that[10] "Biological terrorism affords the possibility of repeated attack, undermining confidence and forcing ever-escalating investments of resources to achieve a modicum of defence." He went on to point out that the terrorist's ability to carry out repeated attacks could remain intact while a government's ability to manage the consequences of the attacks could become exhausted. Here we encounter a new concept of a terrorist *campaign* using bioweapons rather than the usual concept of an isolated attack with biological weapons. This is a much more serious problem that Danzig wishes us to consider. He advises, "[p]lan to defend against a campaign not just an attack".

But just how does a government go about planning to defend against such complexities – so many agents, so many targets, so many different scenarios? In Danzig's view, one way forward is to try to devise a representative range of possible attacks (planning cases) and to work out what capabilities would be required to deal with them. Though the planning cases have to be drawn up with care, he views the cases themselves as much less important than the process of trying to work out what capabilities would be required to deal with them. They are, in short, "an anvil against which to hammer out our hypotheses ... and test the validity of different strategies".

Working in the context of the United States, but presumably considering the thinking of experts in other countries, Danzig suggested four planning cases that he felt would "represent our most significant risks, illuminate how our systems would be taxed, and stimulate a broad range of preparations". The cases were:

1. A large-scale outdoor aerosol anthrax attack.
2. A large-scale outdoor aerosol smallpox attack.
3. An attack that disseminates botulinum toxin in cold drinks.
4. An attack that spreads foot-and-mouth disease among cattle, sheep and pigs.

He argued that these cases will encompass most others we can imagine. For example, plague is a bacterium like anthrax and is contagious like smallpox, so if we have the general capabilities required to deal with anthrax and smallpox we shall have gone a long way towards being able to deal with plague. An attack with plague is thus an included case because it is far less contagious than smallpox and more responsive to treatment than anthrax (although it will require specific vaccines and treatment regimes).

In this chapter a range of cases as suggested by Danzig has been mentioned, but he makes another suggestion we must also note. As biotechnology evolves and the strategic situation develops, we have to be sensitive to the possibility that bioterrorism will also evolve. Danzig suggested the need for a "Case 5 Committee" deliberately tasked with analysing the evolution of the potential threat and proposing changes to the four planning cases when that becomes necessary.

## conclusions

Since the middle of the last century it has been clear that, if sufficient resources are applied to the problem, biological weapons of mass destruction can be created. It has also been clear that lesser antipersonnel attacks on military or civilian targets are possible without the application of such large resources and that agriculture is particularly vulnerable to attack with biological agents. We have been fortunate that to date such resources have been applied only to a limited extent in major hostilities.

It is often forgotten that the Japanese undertook major campaigns in attacking the Chinese with bioweapons before and during the Second World War, and that the United States carried out a large-scale campaign of plant destruction during the Vietnam War using material such as Agent Orange – a synthetic bioregulator. In a sense, then, the growing anxiety about military, and particularly terrorist, use of biological weapons is just the reawakening of public concern about issues that were well aired many decades ago. In reality, very little has changed so far except the renewed perception of threat. However, as we have seen, the revolution in biology has greatly accelerated over the intervening years and we face the threat in coming decades of a much more systematic application of the new biology to hostile purposes. Though today it remains almost certainly the case that carrying out an aerosolized WMD bioattack is only possible via a state programme, in the future – if we are unable to prevent the thoroughgoing militarization of biology – it seems likely that sub-state groups, and perhaps even deranged individuals, may gain the capabilities to cause great harm.

## references

1. A variety of sources have been used for this chapter. In addition to those mentioned in the text, information about attacks on crops have come from Whitby, S.M. (2002) *Biological Warfare against Crops.* Basingstoke, England: Palgrave.
2. Secretary General (1969) *Chemical and Bacteriological (Biological) Weapons and the Effects of their Possible Use.* New York: United Nations.
3. Dando, M.R. (1994) *Biological Warfare in the 21st Century: Biotechnology and the Proliferation of Biological Weapons.* London: Brassey's.
4. World Health Organization (1970) *Health Aspects of Chemical and Biological Weapons.* Geneva.
5. Stricklett, R.D. (1986) *Current Factors Affecting the Possible Use of Biological Weapons by Terrorists.* Technical Note DPG-TA-86-03. Technical Analysis and Information Office, Dugway Proving Ground, US Army.
6. Herum, A.T. (1975) *Covert Biological Weapons Literature Review: Final Report.* DPG-FR-C425A, Dugway Proving Ground, US Army.
7. Leitenberg, M. (2003) Biological weapons and "bioterrorism" in the first years of the 21st century. In *Possible Use of Biological Weapons: Scientific, Legal and International Implications*, pp. 28–95. Como, Italy: ICGEB and Landau Network.
8. Wheelis, M., Madden, L.V. and Cassagrande, R. (2002) Biological attacks on agriculture: low tech, high impact bioterrorism. *Bio-Science*, **52**, 569–76.

9. Shea, D.A. and Grotton, F. (2004) *Small-Scale Terrorist Attacks Using Chemical and Biological Agents: An Assessment Framework and Preliminary Comparisons.* Washington: Congressional Research Service.
10. Danzig, R. (2003) *Catastrophic Bioterrorism – What Is to Be Done?* Washington: Center for Technology and National Security Policy.

chapter eight

# the web of prevention

## introduction

Excepting the possibility of unbridled use of force by some dominant power or world organization, there is only one way to prevent the initiation and maintenance of offensive biological weapons programmes today and in the future. This is by international – multilateral – agreement on an integrated set of policies that will persuade those contemplating the misuse of modern biology for hostile purposes that the effort and costs are just not worth it. The set of policies has been called the "web of prevention" by the International Committee of the Red Cross (ICRC)[1] and ranges from intelligence activities through to the development of better medical treatments:

- good intelligence on proliferation
- strong export controls on critical equipment and materials
- universal international arms control agreements effectively implemented
- associated national legislation on biosafety and biosecurity
- biodefence capabilities in detection, protection and treatment
- clear international agreement to maintain the prohibition
- steady intelligent development of the web to meet the changing nature of the problem.

The ICRC relates this web of prevention to its appeal to us all to help prevent the hostile use of new biotechnology discoveries (Appendix 1). The intention in this chapter is to explore the individual policy elements in more detail, but before doing that it is necessary to clarify what we are trying to prevent.

## what do we want to stop?

There are two clear dangers. First, we know that the major advances in biowarfare in the past century were made in large-scale state programmes. We need to stop such programmes. Second, we suspect that it will become more attractive and easier for sub-state groups (terrorists) to develop more limited, but still very powerful, biological weapons capabilities in the future. We also need to be able to prevent such activities if we are to live in a reasonably secure and civilized world.

There is a relationship between the policies required to deal with the two dangers. Better intelligence, for example, could relate to either state or sub-state programmes, and better medical treatment could be useful to deal with outbreaks initiated by either state or terrorist activities. Furthermore, by preventing state programmes we limit the danger of spill-over of people and materials from such programmes to terrorists. Nevertheless, national legislation and national police activities are the primary means of dealing with the illegal activities of sub-state groups. So, although the web of prevention helps us deal with terrorism, its main focus is on the big problem: that is, preventing major state-level offensive biological weapons programmes. If we do not stop such programmes, we shall end up with the application of a major new technology to warfare. This could transform warfare and create new and dangerous threats to human rights and dignity.

Although it is not a simple matter to prevent a state taking up an offensive biological weapons programme, and though we shall have to pay attention to the need for policy development as the technology develops this century, it should not be assumed that the preventive task of upholding the prohibition of biological warfare is impossible. The historical evidence shows that to achieve a militarily significant offensive biological weapons capability is no easy task. The military have to be cautious and to have assurance that weapons are effective. They need to have several different agent types for different purposes. They might, for example, require both strategic and tactical weapons and both lethal and incapacitating agents. As we shall see in Chapter 9, the amount of agent required for a single attack is not an insignificant quantity, and the military is unlikely to be satisfied with having just one such quantity of a weaponized agent available. Even if they do not require storage of large quantities of agent, they need to know that the capability for producing such quantities on demand is available. Field testing of weapons is essential, plans of operations have to be agreed and troops have to be trained to use the weapons in the operations envisaged (see Figure 7.3). In short, it is no easy matter to hide a militarily significant programme in today's world. The details of what is underway in a country of concern may not be easy to determine, but that something of concern is happening is difficult to conceal. Additionally, as we shall see, though choking off a programme is not as simple as denying fissile material to a nuclear programme, there are points at which the international community can act to greatly complicate, if not entirely frustrate, the proliferator's task.

The bad news is that total prevention may be impossible. The good news is that prevention can be greatly enhanced by the international community acting in concert to strengthen the integrated set of policies, the web of prevention. There have been, and certainly will be, frustrations when new policy developments are slow to be agreed, but progress in any one area can assist in reinforcing the total web. We should therefore act sensibly whenever and wherever the opportunity arises. It is also necessary to be sensitive to the potential negative effects of over-reaction. Fearing inadvertent outcomes of experiments, for example, we should not place such restrictions on scientists that we cut off the flow of potentially useful knowledge directed to benign purposes such as curing cancer and we should not rush to enact such strict anti-terrorist legislation that it threatens the very civil liberties that we wish to protect.

Bearing all that in mind, we turn now to the individual policy elements of the web of prevention.

## intelligence

With the recent intelligence failures in Iraq many will naturally question the place of intelligence operations in the web of prevention. Yet a moment's reflection will bring to mind many such failures in the past, for example Pearl Harbor, Operation Barbarossa (the German invasion of Russia), the fall of France in the Second World War and, particularly in the present context, Allied overestimation of German biological weapons capabilities in that war. Intelligence failures are nothing new and will certainly occur again in the future.

Those failures, however, have not led to the abandonment of attempts to understand better what is happening in a potentially hostile world. What is needed is well-grounded professional analysis and a clear understanding of the difficulties and limitations by those who use the intelligence information. In particular, it must be understood that intelligence analyses are necessarily cautious because of the dangers of missing something of critical importance and are thus likely to tend towards worst-case analyses. Nevertheless, the foundation for good policy has to be the best possible understanding of the external world and much of that will have to come from the best possible intelligence wisely interpreted.

## export controls

In order to have an effective offensive biological weapons capability today a proliferator would have to be able to successfully complete a sequence of activities. The proliferator would have to be able to obtain the pathogenic organisms required. These organisms would have to be grown and concentrated and then processed correctly so that they could be aerosolized or otherwise dispersed. While these operations were carried out, the workers would have to be properly protected. Furthermore, if the proliferator wished to genetically modify an agent to increase its utility, access to the appropriate equipment and techniques used in modern biology would also be required.

Much of the equipment required by a proliferator would be widely available, but certain elements of it are not so easily obtainable or easily replaced by other technologies. One way, therefore, to make proliferation more difficult is to restrict access to such items for countries that have not been able to convince the international community that they have no intention of operating an offensive programme. This is the objective of the Australia Group of countries.[2] The group has been criticized for restricting access to relevant technologies required for peaceful development by some, but it is increasingly obvious that all countries – developed and developing – have a common interest in doing what they can to prevent the proliferation of offensive biological weapons programmes. The Australia Group may be an imperfect mechanism at present, but it is certain that the control of key exports will be an important element in the web of prevention for many years to come. A more relevant question is how to improve and broaden the scope of the export controls in future decades.

The Australia Group began its operations when it was realized how easily Iraq had obtained the necessary elements for its offensive chemical weapons programme – and how those weapons had then been extensively used by Iraq against Iran and against its own citizens in the 1980s. At a later stage, controls were extended to cover the critical elements of an offensive biological weapons programme. At the present time, the Australia Group of countries has agreed to place restrictions on the export of seven categories of equipment and related technologies (Table 8.1) and on listed human, plant and animal pathogens and toxins (Table 8.2).

**Table 8.1** Australia Group equipment list categories*

| | |
|---|---|
| I. Equipment | 1. Complete containment facilities at P3 or P4 containment level |
| | 2. Fermenters |
| | 3. Centrifugal separators |
| | 4. Cross (tangential) flow filtration equipment |
| | 5. Freeze-drying equipment |
| | 6. Protective and containment equipment |
| | 7. Aerosol inhalation chambers |
| II. Related technology | |

* From <www.australiagroup.net>.

It can be seen from Table 8.1, for example, that a request for export of a wide variety of equipment of relevance to an offensive programme would be subject to careful government export controls in an Australia Group country. Items 1 and 6 in the list given in Table 8.1 cover the containment required to work safely with pathogens, items 2, 3 and 4 are concerned with the growth and concentration of the agents and items 5 and 7 with agent preparation and aerosolization. The lists of agents given in the various categories and sub-categories of Table 8.2 are also extensive and include the kinds of pathogens and toxins we have discussed in previous chapters. It will be noticed, in particular, that there are agents listed in regard to attacks on plants and animals in addition to agents relevant just to humans.

Again the message here is that we should not despair of erecting an effective web of prevention. It may not be possible to stop every proliferator, but states would surely not have expended the efforts

**Table 8.2** Australia Group agent list categories*

| | |
|---|---|
| Core list (human pathogens and toxins) | Viruses<br>Rickettsiae<br>Bacteria<br>Toxins (and sub-units)<br>Genetic elements and genetically modified organisms |
| Core list (plant pathogens) | Bacteria<br>Fungi<br>Viruses<br>Genetic elements and genetically modified organisms |
| Core list (animal pathogens) | Viruses<br>Bacteria<br>Genetic elements and genetically modified organisms |

* From <www.australiagroup.net>.

they have on these export controls if they did not believe they could at least place major obstructions in the way of proliferation.

## arms control

As we shall see in more detail in the next chapter, the central Biological and Toxin Weapons Convention was negotiated in the middle of the cold war and agreed without any effective verification system. Unfortunately, efforts during the 1990s to strengthen the BTWC through agreement of a verification protocol eventually failed in 2001 because of opposition from the United States.

Nevertheless, the text of the intended protocol had advanced to such a stage during the years of negotiation that it is possible to see how the negotiators were trying to establish confidence that the parties to the BTWC were living up to their obligations.[3] The Articles listed in Table 8.3 show that the intended protocol was a long and complex document reflecting the various concerns of the states parties.

**Table 8.3** The Proposed Protocol to the Convention on the Prohibition of the Development, Production and Stockpiling of Bacteriological (Biological) and Toxin Weapons and on Their Destruction

**Table 8.3** *contd.*



* From <www.opbw.org>.

Of particular interest to us here is the verification mechanism embedded in the following articles:

Article 4 Declarations
Article 5 Measures to Ensure the Submission of Declarations
Article 6 Follow-up after Submission of Declarations
Article 8 Consultation, Clarification and Cooperation
Article 9 Investigations

and also in the

Annex on Lists (Annex A);
A. Lists of Agents and Toxins
B. Lists of Equipment

Confidence was intended to be steadily increased by the submission of declarations, checking of those declarations through visits from the Technical Secretariat of the organization to be set up for that purpose, clarification mechanisms for issues of concern and mandatory investigations if concerns about what was declared (or had not been declared) could be resolved in no other way.

The total verification mechanism was complex and it is not necessary for our purposes here to review it completely. What is of interest, however, is what was to be declared and subject to checking. An initial declaration was to be required when the protocol came into force and then declarations were to be submitted annually. Increasingly severe penalties for the non-submission of declarations would have ensured that non-submission was unlikely as an option (Article 5). The categories covered by these annual declarations were:

- national biological defence programmes and/or activities against bacteriological (biological) and toxin weapons conducted during the previous year (that is, biodefence);
- maximum biological containment;
- high biological containment;
- plant pathogen containment;
- work with listed agents and/or toxins, e.g. intentional aerosolization of any agent and/or toxin listed in Annex A;
- production facilities.

Within each of these categories further text then specified precisely what elements would have to be declared. For example, whereas the top-level maximum biological containment facilities (of which there are few) would all have had to be declared, high biological containment facilities would only have had to be declared if they were associated with activities such as vaccine production. Vaccine production usually requires, first, production of the agent against which the vaccine is to be used, so vaccine production is obviously a relevant concern. The list of agents in the annex again covered humans, animals and plants and Article 3 made it clear that the list was "not exhaustive". It pointed out that other agents could also be of concern and provided a set of criteria by which assessments could be made of whether to add new agents to the list.

It was a severe setback for the ideal of erecting an effective web of prevention when the negotiations for the protocol failed, but in the proposed protocol we can see the application of the same general strategy as that employed in the Australia Group export controls. Certain facilities and activities are crucial in an offensive programme and many of these are potentially dual-use, so, by checking that they were not being misused, confidence in compliance could have been much enhanced.

Two further points about the checking of declarations should be noted. Firstly, the number of visits to check declarations was intended to be limited so as not to impose too great a burden on any one country or industry. So it could be asked why reliance should be placed on declarations that might well *not* be checked. The answer can be found by thinking about income tax returns. If you have to fill in a return, and you know that only one in ten will be checked you might decide to cheat. Yours, however, might be the very one that is checked, so cheating is a risky option. Cheating, in fact, is limited in two ways. If you sign up to a system, you are unlikely to cheat in the first place, and a low level of checking is all that is required to encourage most to conform. Secondly, an evasion could be envisaged where a proliferator might decide not to declare a site where proliferant activities were taking place. The proliferator now has an even bigger problem. Instead of declaring a facility, but attempting to hide the illegal activity, the proliferator has to hide both the activity and the facility. Article 8 allows for questions to be asked and Article 9 allows for investigation of concerns that cannot be resolved. So a proliferator has the ugly possibility of having a non-declared facility both detected and investigated.

Finally, it might be argued that a genuine proliferator would avoid the verification mechanism and all such hassle by not joining the protocol or even by remaining outside the BTWC itself. That would be an option, but with most states now signed up to the BTWC, and if agreement had been reached then also subject to the protocol, such an isolated state would have great difficulty persuading others of its good intentions. All other aspects of the web of prevention would therefore be applied in order to make proliferation as difficult as possible for that state.

The net result of this analysis is to suggest that the BTWC will have to be reinforced at the international level with something akin to, or part of, the proposed protocol as quickly as international (and some national) conditions allow. In the meantime, attention can be given to national implementation. Article IV of the BTWC requires that effective national measures are taken to implement the prohibition. Some states did this long ago, and recently some states have reviewed and reinforced their national measures. Yet we know that the measures taken by many states are inadequate and other states have taken none at all. This issue was one of the topics of the interim process agreed at the 2001–2 Fifth Review Conference of the BTWC (see Chapter 9) and, since effective national measures (and international co-operation) are essential for dealing with sub-state terrorist activities, it is greatly to be hoped that the required report at the 2006 BTWC review will demonstrate much progress on improving national measures. Similarly, in regard to the interim process discussions on biosecurity, it is in all our interests to ensure that biosecurity regulations are adequate to prevent the inadvertent or malign diversion of agents and equipment for hostile purposes.

## biodefence

As is clear from the detailed discussion of major agents of concern in Chapter 5, there is much that can be done to improve our capabilities for defending against attacks with biological weapons. There are obvious gaps in our means of detection, diagnosis, prevention and treatment which can and should be remedied. That is not to argue that we can cover all possible contingencies; but each improvement we can realistically make introduces further complications for, and so helps to deter, those who might consider such attacks. More broadly, as the terrorist threat has been seen to increase, governments have had to reorganize in order to understand the risks better, to attempt better prevention of attacks and to deal with the consequences if one should succeed. The most noticeable reorganization has been in the United States with the creation of the Department of Homeland Security, but other countries such as the UK have made significant changes to national and regional systems. In this context, an attack with biological weapons is considered to be one of the major threats that may arise. So sensible measures of protection, detection and treatment should be taken against likely threats – measures that are both allowed by the BTWC and would be supported by most people.

However, there are dangers if caution is not exercised over biodefence. The history of biological programmes in the twentieth century demonstrates the dangers of misperceptions arising, leading to the initiation of retaliatory programmes on the basis of erroneous intelligence. An example of such a misperception and response was the reaction to the presumed reactivation of the German offensive biological weapons programme in the 1930s (see Chapter 2). Moreover, if biodefence strays too far into threat assessment, and such threat assessment is not understood by other states, there is particular danger. In the United States around the turn of the twenty-first century there were at least three programmes that could have been a cause for concern. One of these programmes involved replicating parts of a Soviet biological weapons munition to test its characteristics. This activity is particularly difficult to reconcile with the second paragraph of Article I of the BTWC's prohibition (see the full text of the BTWC in Appendix 2). As biodefence expenditures increase in many countries it will be important to keep restraint and transparency in mind. The three American programmes of concern here were not even detailed in the Confidence-Building Measures required annually under the BTWC and could therefore have appeared particularly suspicious when revealed.

A final point about biodefence needs to be made. Given the range of potential targets of attack and agents for attack, it is unlikely that *all* possible dangers can be countered. Indeed, there is good reason to believe (Chapter 7) that the advantage will increasingly lie with the attacker if we allow an action/reaction arms race to proceed in coming decades. Biodefence is part of the solution if seen in perspective as part of the web of prevention. However, it could easily become part of the problem if it gets out of control.

## maintaining the prohibition

In present circumstances, after the second Gulf war, mention of international determination to respond to any deviation from the norm, and break-out from the prohibition embodied in the BTWC, is likely to evoke uneasy images of faulty intelligence and pre-emptive wars. However, pre-emptive war is not the only method of suppressing the activities of a likely proliferator. In the case of Iraq it would now appear that the international sanctions placed on its trade and activities by the international community in 1991, and developed through the next decade, were part of the reason for the closure of Saddam Hussein's offensive biological weapons programme.

The ongoing verification and monitoring systems imposed on Iraq were a particularly severe form of intrusion and are unlikely to be replicated. Yet Security Council Resolution S/22871/Res 1 of 2 October 1991 also stated that the efficacy of the monitoring and verification provisions "would be enhanced if they were complemented by transparency and timely information as regards any future sale or supply by other States to Iraq of relevant items with dual use". The resolution continued:

> Such a comprehensive approach would call for the development of a mechanism that:
>
> (a) Upholds the prohibition on the sale and supply to Iraq by other States of any weapons or related items prohibited ...

(b) Provides for timely information about any sale or supply to Iraq by other States of items that could be used not only for permitted purposes but also for purposes prohibited.

By 1995, in Resolution S/1995/208, the Security Council had refined the items that it wanted to be subject to such control.[4] The items contained in the relevant appendix of the resolution are summarized in Table 8.4. Again the particular types of agents were

**Table 8.4** Summary of items to be reported*

1. Micro-organisms, toxins, other organisms and genetic material (as listed separately)
2. Biohazard containment and decontamination items
3. Fermentation equipment
4. Equipment useable for processing, handling, transporting or storing micro-organisms or their products or components.
5. Formulated powdered complex media or concentrated liquid complex media for growth of micro-organisms
6. Detection and assay systems for micro-organisms, toxins or genetic material and specially designed reagents (for the separately listed agents)
7. Equipment and reagents for use in molecular biology research and specially designed components thereof
8. Equipment capable of dispersing aerosols and specially designed components thereof
9. Equipment useable in the study of aerosols and specially designed components thereof
10. Equipment designed for the microencapsulation of living organisms or their products or components, including toxins or other biological material
11. Vaccines for the micro-organisms or toxins (listed separately)
12. Documents, information, software or technology for the design, development, use, storage, manufacture, maintenance or support of entries 1–11 listed above
13. Munitions, rocket or missile warheads capable of disseminating biological warfare agents

* From Security Council, op. cit.

given in a separate listing and the equipment relevant under each of the items was stated under each item. So, for instance, the specially designed (HEPA) filters required for biological containment are listed as point 2.4 under item 2 and aerodynamic particle-sizing equipment is point 9.3 under equipment suitable for the study of aerosols.

By 1995 the Security Council was undertaking very detailed control of anything that might be relevant to an offensive programme in Iraq. Such detailed control would obviously not have been acceptable to the states parties negotiating the BTWC protocol and would have overloaded the Australia Group's efforts to deal with many more than one country's imports. Yet the same general pattern emerges. It is possible to think through the requirements of an offensive programme and to identify elements that can be monitored, in order to increase confidence in compliance, or subjected to controls to prevent proliferation. The international community is not bereft of options.

## building the regime

The same overall message comes through if we think carefully about what to do now and in the future to strengthen and build up the web of prevention. It is all too easy to look at the difficulties encountered in recent years and to conclude that there is little that can be done. The opposite is true. From the start, part of the problem has been sheer ignorance about the biological threat and what has been done in the past to deal with it – understandable given fifty years of concentrating on the nuclear menace. That is changing as more and better-organized information becomes available to the general public and the media. But there is a problem here in that, unlike the international community of physical scientists, biologists have little tradition of taking an interest in security issues. Therefore the people who should really be able to give the best advice to the public and to politicians are still hardly aware of their grave responsibilities. A key issue for civil society is how to make the whole life sciences community around the world more aware and willing to bring their expertise to bear on the issue of preventing the misuse of their science.

Traditionally, there have also been few non-governmental organizations with an intense long-term interest in biological arms control issues. This is changing with the creation of a Geneva-based co-ordinating Bioweapons Prevention Project (BPP) whose aim is to assist more and more non-governmental organizations to work in this area. The BPP is laying particular emphasis on helping organizations from the developing world to make their interests and activities more widely known and understood.

At an entirely different level, the United Nations Security Council is increasingly seeing the need for urgent co-ordinated action to prevent the proliferation and potential misuse of biological agents. Increased fears of terrorist use of weapons of mass destruction have led to the consideration of what further actions might be taken against the possible use of *all* such weaponry by terrorists – nuclear, chemical and biological. Thus United Nations Security Council Resolution 1540 requires all states to refrain from providing any form of support to non-state actors seeking any weapons of mass destruction, to enact and enforce effective national laws to that end and to report regularly on what they have done. Recent experience of the Chemical Weapons Convention shows why an emphasis on national legislation is required. Despite the clear need for national legislation to bring the prohibition into real effect, and the efforts of the Organization for the Prohibition of Chemical Weapons (OPCW) to provide assistance, it has not been easy in many states to enact effective legislation. There are, however, concerns about the approach that may lie behind Resolution 1540. The OPCW specifically rejected a name-and-shame approach to improving the states parties' performance. Instead, a multifaceted implementation plan is providing a variety of different forms of co-operative assistance, and achieving good results.

The implementation of Security Council Resolution 1540 could similarly be done with a co-operative approach to help states to tighten up their legislation while acknowledging their real difficulties in lack of resources. There are fears, though, that the intention may be to become much more confrontational, an approach that could create a great deal of friction. Another worry is that measures such as Resolution 1540 might undermine the centrality of the multilaterally agreed BTWC, the core of the prohibition regime – particularly since it has not been possible to persuade the United States of the need for the convention to have a verification protocol.

It should not be thought that such sweeping responses can only be conceived by organizations like the Security Council. The

Harvard/Sussex Universities academic group, which has long played a leading role in the study of chemical and biological weapons, for example has argued for and carried out considerable development of an international treaty to criminalize any breaking of the prohibitions embodied in the BTWC and the CWC. Such a treaty, if implemented by the international community, would allow no exceptions. Any person, from heads of state downwards, would be treated as an international criminal if they broke the prohibitions. Also, as we saw at the beginning of this chapter, the International Committee of the Red Cross and its national societies have taken considerable interest in the issue – particularly in the possible ways in which future technologies may be misused – and have argued strongly for a tough web of prevention to be developed and maintained.

## conclusions

There is a danger that by raising the difficult issue of preventing the hostile use of biology it is inadvertently suggested that there is little that anyone can do to counter the threat. Such pessimism may be reinforced when the difficulties of reaching international agreements are also considered. The overall intention of this chapter has been to argue that such pessimism is unjustified, indeed that the very opposite is true – that there are many ways in which individuals and groups can take effective action at many different levels. Furthermore, action that strengthens the web of prevention at one level helps to reinforce all of the other elements in the web because the aim is to make it as difficult as possible for anyone to consider the hostile use of biological agents.

Much has already been done to construct the web of policies we need, but there is still much more to do. There are particular difficulties if the central norm-creating international agreements are weak. It is to the BTWC regime itself that we turn in the next chapter.

## references

1. ICRC (2002) *Biotechnology, Weapons and Humanity: Summary Report of an Informal Meeting of Government and Independent Experts.* Geneva.
2. For details of the Australia Group and its export control system see <www.australiagroup.net>.

3. For the final version of the negotiation text see BWC/AD HOC GROUP/56–2, <www.opbw.org>.
4. Security Council (1995) *Plan for Future Monitoring and Verification of Iraq's Compliance with the Relevant Parts of Section C of Security Council Resolution 687 (1991)*. S/1995/208. United Nations, 17 March.

# the failure of arms control

## introduction

As the application of science and technology to warfare accelerated during the industrial revolution the lethality of the weapons systems, and the casualty and death rates, increased alarmingly. There were significant innovations such as the use of railways to move large numbers of men and equipment, for example in the American Civil War in the mid-nineteenth-century. Perhaps even more significant than the railways, however, was the smooth-bore musket's replacement by much more accurate rifles that rendered previous infantry tactics useless. If infantry charges were attempted, the new rifles led to large-scale slaughter of the attackers. These developments did not go unnoticed outside the military, and efforts were made by governments and diplomats to find means of limiting the use of at least some weaponry. Thus a meeting of government representatives led to the St Petersburg Declaration outlawing dumdum bullets (blunt and therefore very destructive) as early as 1868.

Among the notable scientific and technological developments of the later nineteenth century, chemistry stands out – particularly in Germany. A series of scientific developments led to rapid and hugely profitable innovations such as colour dyes for cloth. The possibility that the new developments in chemistry could find hostile applications was appreciated and some restrictions were agreed at the conferences in The Hague of 1899 and 1907. However, these agreements, and the long-held abhorrence of the use of poisons in warfare, did not prove sufficient to prevent the widespread use of chemical weapons by both sides during the First World War of 1914–18.

The terrible deaths and long-term injuries caused by chemical weapons during the war led to considerable efforts to outlaw such weapons in the interwar years. These efforts came to fruition in the 1925 Geneva Protocol, which is now widely regarded as customary international law, binding on all states. The control regime developed by the international community to restrict any particular sphere of activity can be regarded as comprising the rules embodied in the relevant agreements, the organizations set up to enforce the rules and the actual enactment of the rules in state practice. Despite its limitations, the 1925 Geneva Protocol is the foundation of the biological and toxin weapons regime and the chemical weapons regime.[1]

## the 1925 Geneva protocol

As noted in Chapter 2, the protocol consists of a short statement, a declaration of the purpose of the agreement and then five paragraphs dealing with accession, language and the like. The opening statement is, in effect, a preamble that links the protocol to previous developments in international law:

> Whereas the use in war of asphyxiating, poisonous or other gases, and of all analogous liquids, materials or devices, has been justly condemned by the general opinion of the civilized world; and
>
> Whereas the prohibition of such use has been declared in Treaties to which the majority of Powers of the World are Parties; and
>
> To the end that this prohibition shall be universally accepted as part of International Law, binding alike the conscience and the practice of nations.

The declaration continues as follows:

> That the High Contracting Parties, so far as they are not already Parties to Treaties prohibiting such use, accept this prohibition, *agree to extend this prohibition to the use of bacteriological methods of warfare* and agree to be bound as between themselves according to the terms of this declaration. (Emphasis added)

The negotiations, as we have seen, were at first concerned with restricting chemical weapons only, but the Polish delegation argued that bacteriological weapons could be just as devastating and it was agreed to extend the agreement to cover such weapons. At that time, the nature of viruses and viral diseases was not well understood, hence the use of the term "bacteriological warfare". However, that wording is now taken to cover *all* biological agents.

The 1925 Geneva Protocol, nevertheless, has many deficiencies. It is clear from the text that it only prohibits *use*. The states parties such as the UK which developed and produced biological weapons for retaliatory purposes during the Second World War were not therefore in contravention of the protocol. Additionally, the text refers specifically to use "in war" but does not define what that means. The text also makes clear that the prohibitions apply only to fellow parties to the protocol. States also registered a variety of reservations when they joined the agreement. Crucially, many supplemented the restriction to fellow parties by extending it to allies of fellow parties. Thus a state party to the agreement whose ally used a prohibited weapon would not itself be safe from retaliatory use by an affected fellow party. In short, many commentators regarded the 1925 Geneva Protocol as a "no-first-use" agreement. However, over the last eighty years most of the reservations to the protocol have been removed. This has strengthened the core of the regime, but the process needs to be completed through *all* reservations being lifted. There are still a few states that have yet to become parties to the protocol and this deficiency also needs to be remedied.

It is widely agreed, and fortunate, that biological weapons have only been used on a large scale by one country, Japan, over the last seventy years. This attempt to use biological warfare was carried out in an unsophisticated manner in China during the Second World War. Chemical weapons, though available in large quantities to both the Allied and Axis powers, were not used in that war. A number of reasons have been suggested for this non-use: military distaste for the weapons; public and political opinion; fear of retaliation; and the force of the 1925 Geneva Protocol.

As we saw in Chapter 4, the victorious powers came out of the war with intact biological weapons programmes and large stocks of chemical weapons. It was subsequently found that Germany had discovered the much more lethal nerve gases and had developed them as chemical weapons agents. In the early deliberations of the United Nations after the war it was recognized that chemical and biological weapons were best viewed along with the new atomic (nuclear) weapons as belonging to a special class of weapons capable of mass destruction. The removal of such weapons from the arsenals of states was one of the subjects of the more general deliberations on "General and Complete Disarmament" which followed. No great progress had been made on this topic by the end of the 1960s and attention had by then turned to making progress on more limited problems. One of these was the problem of chemical and biological weapons: the experts' report for the UN Secretary General in 1969 (see Chapter 7) re-emphasized their dangers and strongly urged an agreement to eliminate these weapons. The United Kingdom then took the lead in putting forward a text for what eventually became the 1975 Biological and Toxin Weapons Convention (BTWC).

## the biological and toxin weapons convention

The British text proposed dealing separately with biological and chemical weapons. In comparison with what was eventually agreed, the British draft was quite tough. It proposed first to supplement the 1925 Geneva Protocol's prohibition on use, stating:

> Each of the parties to the Convention undertakes, insofar as it may not already be committed in this respect under Treaties or other instruments in force prohibiting the use of chemical and biological methods of warfare, never in any circumstances ... to engage in biological methods of warfare.

This proposal, regarded by some as perhaps undermining rather than reinforcing the protocol, did not receive wide support. Perhaps more importantly for the future, the UK's suggestion that the agreement also ban research did not survive the negotiations.

Initially, the British suggestion that negotiations on chemical and biological weapons should be divided did not meet with widespread approval. Many regarded it as a ploy to help the United States through its difficulties with world public opinion over the use of tear gas and herbicides in Vietnam. Eventually, and for reasons that remain unclear, the Soviet Union gave support to the idea and with the United States pushed negotiations forward to arrive at the text of the convention that we have today. Compared with the post-cold-war 1997 Chemical Weapons Convention, the mid-cold-war

BTWC is a very slight document. The full text of the 15 articles is given in Appendix 2 and the provisions of the articles can be summarized as follows:

1. Not to develop, produce, stockpile or acquire agents, weapons etc.
2. To destroy stocks.
3. Not to transfer to or assist others.
4. To take national measures.
5. To consult and co-operate in solving problems.
6. Option to lodge complaint with the Security Council.
7. To provide assistance in the event of a violation.
8. No detraction from the Geneva Protocol.
9. Obligation to continue negotiations on chemical weapons.
10. Co-operation for peaceful purposes.
11. Amendment.
12. Review.
13. Duration.
14. Signature, ratification, accession and deposit.
15. Languages.

The convention was opened for signature in 1972 and entered into force in 1975. Some 150 states are now parties – including the vast majority of the more influential states. The BTWC can therefore be seen as a significant addition to the 1925 Geneva Protocol.

The crucial Article I of the Convention declares:

> Each State Party to this Convention undertakes never in any circumstances to develop, produce, stockpile or otherwise acquire or retain:
>
> 1. Microbial or other biological agents, or toxins whatever their origin or method of production, *of types or in quantities that have no justification for prophylactic, protective or other peaceful purposes.*
> 2. Weapons, equipment or other means of delivery designed to use such agents or toxins for hostile purposes or in armed conflict. (Emphasis added)

So the prohibition applies to five activities – development, production, stockpiling, acquisition and retention. The British ideas of prohibiting research and use had been lost in the negotiations. The prohibitions relate to any microbial or other biological agents or toxins "of types and in quantities that have no justification for prophylactic, protective or other peaceful purposes". This has become known as the "General Purpose Criterion" and is clearly intended to apply for all time. The prohibition is absolutely sweeping and at the time was hailed as the first disarmament agreement that eliminated an entire class of dangerous weaponry.

The implications of Article I led to the second article, which provides for the rapid destruction or diversion to peaceful purposes of any agents held by a state party following the convention's entry into force. Article III prevents transfers, assistance, encouragement and inducement and Article IV requires states to take necessary measures to implement the convention nationally. The UK, for example, introduced a draconian law for this purpose. Article V then provides for consultations in solving problems that may arise and Article VI for states parties to take complaints about breaches of the convention to the Security Council.

Article VII provides for support of a party subject to a violation of the convention and Article VIII emphasizes that nothing in the agreement detracts from the obligations undertaken by any state party to the 1925 Geneva Protocol. In Article IX the parties undertake to continue negotiations on a prohibition of chemical weapons – an obligation met some twenty years later. Then in Article X – somewhat in contradiction to Article III – it is agreed that:

> 1. The States Parties to this Convention undertake to facilitate, and have the right to participate in, the fullest possible exchange of equipment, materials and scientific and technological information for the use of bacteriological (biological) agents and toxins for peaceful purposes.

Given the dual-use nature of much of the equipment, materials and information in this field of science and technology, the friction over different interpretations of Articles III and X which has followed is hardly surprising.

Article XI provides for potential amendment of the convention, but states parties are loath to undertake this for fear of the whole set of agreements underlying the convention beginning to unravel. More importantly, Article XII provided for the first Five-Year Review Conference in 1980 which mandated the following 1986 review. This process will continue with the Sixth Five-Year Review slated for 2006. The remaining articles of the convention deal with duration (unlimited), language and so forth.

## deficiencies of the convention

The preamble to a convention is not legally binding in the same way as the articles, but it does provide an idea of what the states parties are trying to achieve. The preamble to the BTWC leaves no doubt about the seriousness of intent; its opening paragraph reads:

> The States Parties to this convention, Determined to act with a view to achieving effective progress towards general and complete disarmament, including the prohibition and elimination of all types of weapons of mass destruction, and convinced that the prohibition of the development, production and stockpiling of chemical and bacteriological (biological) weapons and their elimination, through effective measures, will facilitate the achievement of general and complete disarmament under strict and effective international control ...

This point is emphasized in paragraph 7, which reads:

> Convinced of the importance and urgency of eliminating from the arsenal of States, through effective measures, such dangerous weapons of mass destruction as those using chemical or bacteriological (biological) agents ...

Yet we have only to recall that the Soviet Union embarked on a massive offensive biological weapons programme expansion, as the convention came into force, to understand that the measures undertaken were quite inadequate.

France, the depositary state for the 1925 Geneva Protocol, and well aware of the dangers from experience of its own offensive programme, enacted measures nationally but for a long time refused to join the BTWC because of the lack of effective verification. How could the provisions of Articles V and VI for consultation and the possibility of complaint really have been effective in ensuring that the states party to the convention were living up to their obligations?

A further serious deficiency, which was to become more and more evident in later years, was that no organization had been put in place to take care of the convention, its effective implementation and its development between Review Conferences. By contrast, the CWC has a major international organization – the Organization for the Prohibition of Chemical Weapons – to carry out verification and other activities.

With these major deficiencies in the BTWC, the Five-Year Review Conferences have mandated a series of efforts directed at dealing with the problem of verification – and the lack of confidence and trust that the deficiencies engender. The first such effort was to agree a series of Confidence-Building Measures at the Second Five-Year Review Conference in 1986.

## confidence-building measures

As a newer generation of leaders came to power in the Soviet Union and the fears of the later cold war years diminished, Confidence-Building Measures appeared to provide a general way forward in arms control. These measures were voluntary, politically binding but not legally binding and seemed to provide a low-cost method of increasing trust. In 1986 states parties to the BTWC agreed to implement such measures "in order to prevent or reduce the occurrence of ambiguities, doubts and suspicions, and in order to improve international co-operation in the field of peaceful bacteriological (biological) activities". The measures agreed are as follows:

> Exchange of data on research centres and laboratories that have very high safety standards in handling biological materials that pose a high risk.
>
> Exchange of information on outbreaks of disease and similar occurrences caused by toxins that appear to be different from normal.
>
> Encouragement of publication of results of research directly related to the Convention and promotion of the use of such knowledge for permitted purposes.
>
> Active promotion of contacts between scientists engaged in research directly related to the Convention, including exchanges for joint research.

An experts' meeting followed in 1987 in order to work out the details. Unfortunately, from the very beginning it was clear that these CBMs would do little to help. By 1990 four annual rounds of data submission had taken place, but only thirty-six states had taken part. The situation has not radically changed since then. Very few states have regularly submitted adequate information, despite an agreement in 1991 to improve the set of CBMs in order to enhance their potential for transparency and confidence building (Table 9.1).

**Table 9.1** The improved CBMs agreed in 1991

| Measure | | Description |
|---|---|---|
| Declaration Form | | Nothing to declare or nothing new to declare |
| A | (1) | Exchange of data on research centres and laboratories |
| | (2) | Exchange of information on national biological defence research and development programmes |
| B | | Exchange of information on outbreaks of infectious diseases and similar occurrences caused by toxins |
| C | | Encouragement of publication of results and promotion of use of knowledge |
| D | | Active promotion of contacts |
| E | | Declaration of legislation, regulations and other measures |
| F | | Declaration of past activities in offensive and/or defensive biological research and development programmes |
| G | | Declaration of vaccine production facilities |

With that dismal record, many states parties at the 1991 Third Review Conference wanted to move to a much tougher system. What was eventually agreed, however, was to investigate the possibilities for verification in a process that became known as "VEREX".

## VEREX

The instructions to the experts who met four times in 1992 and 1993 were that they should "identify and examine potential verification measures from a scientific and technical standpoint". The measures were clearly related to the central Article I of the BTWC, since they were to determine:

> Whether a State Party is developing, producing, stockpiling, acquiring or retaining microbial or other biological agents or toxins, of types and in quantities that have no justification for prophylactic, protective or other peaceful purposes.

The measures could be assessed singly or in combination and were to be judged against this defined set of criteria:

> Their strengths and weaknesses based on, but not limited to the amount and quality of information they provide, and fail to provide;
>
> Their ability to differentiate between prohibited and permitted activities;
>
> Their ability to resolve ambiguities about compliance;
>
> Their technology, material, manpower and equipment requirements;
>
> Their financial, legal, safety and organisational implications;
>
> Their impact on scientific research, scientific co-operation, industrial development and other permitted activities, and their implications for the confidentiality of commercial proprietary information.

The work of the VEREX group was restricted because what was done could be stated in a report but the implications of what was agreed could not. However, the mandate from the 1991 review also allowed for the report to be circulated to states parties and "If a majority of State parties ask for the convening of a conference to examine the report ... such a conference will be convened. In such a case the conference shall decide on any further action." The mandate was a compromise between those who wanted to move forward on stronger verification and those – including the powerful United States – who were much more reluctant.

Despite such differences, the experts took a pragmatic approach to their task. A list of twenty-one measures were examined, sometimes in combinations. These were:

- surveillance of publications
- surveillance of legislation
- data on transfers, transfer requests and production
- multilateral information sharing
- exchange visits (off-site)
- declarations
- surveillance by satellite
- surveillance by aircraft
- ground-based surveillance (off-site)
- sampling and identification (off-site)
- observation (off-site)

- auditing (off-site)
- exchange visits – international arrangements
- interviewing (on-site)
- visual inspection (on-site)
- identification of key equipment (on-site)
- auditing (on-site)
- sampling and identification (on-site)
- medical examination (on-site)
- continuous monitoring by instruments (on-site)
- continuous monitoring by personnel (on-site)

The experts also managed to reach some consensus agreements. For example:

> Declarations, if properly structured, could be an important mechanism for building up a picture of the biological activities in a nation ...
>
> On balance, it would appear from this evaluation that declarations have a high status in terms of potential utility.

The final report concluded, in part, that "potential verification measures as identified and evaluated could be useful to varying degrees in enhancing confidence, through increased transparency, that States Parties were fulfilling their obligations" and that "some of the potential verification measures would contribute to strengthening the effectiveness and improve the implementation of the Convention". On this positive outcome a majority of states parties acted to request a Special Conference. This was held in 1994 and mandated the work that led to the ill-fated verification protocol rejected by the United States in 2001.

In discussing verification of the BTWC it is important to be clear about what it is important to verify. The main reason for a verification protocol was never to prevent a lone terrorist developing a biological agent in his or her garage. That is the job of national authorities. What the BTWC verification system set out to do was to help to ensure that no states parties were engaged in offensive biological weapons programmes.

At an open meeting of government and non-government groups in 1996 a British expert set out what would be required to generate a militarily significant amount of anthrax.[2] He concluded that the equipment required would "not be very small" and the proliferators would have to demonstrate convincingly why it was needed for peaceful purposes. In addition, he concluded that for militarily significant proliferation, more than one agent and more than one delivery system would be needed. This would obviously further compound the explanatory difficulties. In short, this expert argued that effective verification of compliance with the BTWC was not impossible.

## negotiations of the BTWC verification protocol

The Special Conference to consider the VEREX report was held in September 1994 and mandated the work of the Ad Hoc Group (AHG) which functioned through to July 2001. The mandate was not to produce politically binding (voluntary) measures; rather, the AHG was charged with agreeing a legally binding instrument. The objective was stated in part as "to consider appropriate measures, including possible verification measures, and draft proposals to strengthen the Convention, to be included, as appropriate, in a *legally-binding instrument*" (emphasis added).

The AHG was instructed to consider a range of issues:

- definition of terms and objective criteria;
- enhanced confidence-building and transparency measures;
- a system of measures to promote compliance;
- specific measures to ensure effective and full implementation of Article X;
- protection of sensitive commercial proprietary information and legitimate national security needs;
- avoidance of any negative impact on scientific research, international cooperation and industrial development.

Consider the fourth item in the list, "Specific measures to ensure effective and full implementation of Article X". As can be seen from Appendix II, this article of the BTWC is concerned with co-operation and economic development. For the developing world, struggling to deal with infectious diseases long removed from the developed world, this is an important part of the multilateral agreement. Its importance has also grown, since the revolution in the life sciences seems likely to generate great dividends in the form of improved protection against, and treatment of, malaria and other deadly diseases.

So the protocol was not confined to dealing with the issue of compliance with the prohibition embodied in Article I of the convention. However, in regard to compliance with Article I the AHG was to consider "A system of measures to promote compliance with the Convention, including as appropriate, measures identified, examined and evaluated in the VEREX Report". Furthermore, these measures "should apply to all relevant facilities and activities, be reliable, cost-effective, non-discriminatory and as non-intrusive as possible, consistent with the effective implementation of the system and should not lead to abuse". In short, the AHG had been handed a considerable task. It also had to achieve its results by consensus. Powerful states were therefore in a position to slow down the pace of negotiations or to prevent final agreement if they wished to do so.

Despite these handicaps and the wide disparities in resources available to the different states parties, this was a genuinely multilateral process. Some fifty states parties were regularly represented at the meetings of the AHG, and many different countries contributed working papers setting out their views on aspects of the matters under discussion. One noticeable problem throughout was the lack of media and public interest in these crucial meetings and, in particular, the very small number of well-informed non-governmental organizations able to track, report and lobby for various solutions to the problems that arose. Some of the negotiators involved took to referring to their work as the "orphaned" negotiation. The world-wide scientific and medical communities unfortunately showed little critical interest, and they must surely bear some of the blame for the subsequent failure to reach agreement.

At the first meeting of the AHG in January 1995 it was agreed that after each meeting a procedural report would be prepared and also that this would have an annex reporting the work of the group. The development of the work can therefore be followed through these detailed reports, all available on the internet.[3] Also at the first meeting, a number of "friends of the Chair" were appointed to assist the Chairman in dealing with the different topics under discussion. Initially there were four such appointments to cover the first four topics listed above, but the number of such friends of the Chair rose as more detailed work was required. This again illustrates the multilateral nature of the negotiations.

The Fourth Five-Year Review Conference of the BTWC in 1996 encouraged the AHG to move forward on its mandate. After an initial exploratory phase building on VEREX, the group transitioned to the development of a "rolling text" of a protocol in July–August 1997. A final phase of detailed negotiations on the key elements of the protocol began in January 1999. Though the AHG met for only a few weeks at a time and meetings were usually scheduled only two or three times each year, nevertheless the rolling text steadily grew. The overall configuration of the developing text could be seen from the document produced in mid-2000, for example. The presence of square brackets indicated disagreement over certain sections – even over the titles of some articles and appendices. Many more brackets remained in the text itself.

It was clear to observers that there was a diversity of interests among the states parties, and it was no easy matter to resolve these. A reasonable idea of the range of contentious issues can be seen from the list produced by one knowledgeable commentator in the late 1990s:

- definitions, lists and criteria;
- declarations;
- challenge inspections of facilities suspected of a treaty violation;
- field investigations of unusual disease outbreaks (possibly associated with the covert use of biological weapons or an accidental leak from a clandestine development or production facility);
- non-challenge visits to declared facilities;
- protection of confidential information;
- scientific and technological co-operation in the peaceful uses of biotechnology;
- the non-transfer of equipment and know-how needed for the production of biological weapons.

One critical difference was over the form of the compliance measures required to ensure that states parties were observing their obligations under Article I of the BTWC. The European Union, for example, was in favour of a strong mechanism built up in the same way as the recently agreed Chemical Weapons Convention. Speaking on behalf of the EU, the Portuguese Ambassador at the nineteenth session of the AHG on 13 March 2000 stated that:

> The EU favours *a comprehensive declaration regime* that will capture in a balanced way all facilities and activities relevant to the Convention ...
>
> The Protocol must contain an *effective mechanism for follow-up of declarations in the form of visits* ...

> The Protocol must contain *appropriate clarification procedures ...*
>
> The Protocol must include provisions for *rapid and effective investigations.*

So the EU was arguing that states parties should make declarations of relevant facilities and activities. Then, as under the CWC, the international organization overseeing the protocol (Organization for the Prohibition of Biological Weapons, OPBW) would make visits to check these declarations. In what was to be an advance on the CWC, before resorting to investigations there would also be provisions for clarification of issues of concern to states parties, but this would be backed up, as in the CWC, by a challenge investigation process if the issue could not be resolved. This system would not necessarily detect every minor non-compliance, but supporters argued that it would form a strong deterrent to non-compliance because of the likelihood of detection. It would therefore form an essential element in a web of policies at different levels which would, in total, persuade potential proliferators that biological weapons were not worth pursuing.

One issue that inevitably arose from this approach was the safety of national security and confidential business information during the visits that would routinely be carried out to check declarations at sites of potential relevance to the BTWC. Obvious examples would be the chemical and biological defence facility at Porton Down in the UK, and any vaccine production facilities – which necessarily have large-scale fermentation capabilities that might be misused relatively easily. A number of states parties had carried out practice visits and challenge inspections during the negotiation of the CWC and followed this up with the same sort of practice visits and inspections at facilities relevant to the BTWC. The reports made to the AHG about these trials strongly suggested that with proper procedures – what was termed "managed access" – it was quite possible to carry out the visits and investigations without undue risk to important information.

This view was not shared by a number of other states parties. In particular, there was vocal opposition from the huge US pharmaceutical industry and its linked trade associations. Most of the negotiations took place during the Clinton presidency in the United States and, though America did not openly oppose the reaching of an agreement and maintained some flexibility in its position, it was not able to take a leadership role in the negotiations. In the absence of strong directives from the top of the administration, the different government agencies in Washington could not reach a compromise position and US allies (for example, in the EU), though they watered down what they really wanted, could never reach the point where the US was satisfied. Meanwhile, there were other states parties that did little to help the negotiations reach a conclusion which were not at all unhappy to see the negotiations drift along inconclusively.

Strong supporters of the idea of strengthening the BTWC through agreement of a verification protocol and an associated organization were not deterred. During the twenty-second AHG session a number of delegations called for the Chairman to put forward his own ideas for a compromise "composite text". The Chairman of the negotiations, Ambassador Tibor Tóth of Hungary (who had also chaired the VEREX meetings), had used a number of different strategies to gradually close the gaps between the different states parties. After the end of the session in March 2001 he therefore produced a complete composite text. This was based firmly on the latest version of the agreed rolling text but had been usefully reordered in some respects in an attempt to bridge the remaining differences between the parties. Supporters of the idea of a protocol would have liked a stronger text but still supported it as the basis for going on to conclude a protocol.

By this time, the new Bush administration was in power in Washington. There was now no lack of leadership from the top. The US Ambassador in Geneva told the AHG in July that the United States would not accept the protocol text and further that

> we are forced to conclude that the mechanisms envisioned for the Protocol would not achieve their objectives, *that no modification of them would allow them to achieve their objectives*, and that trying to do more would simply raise the risk to legitimate United States activities. (Emphasis added)

In short, the US rejected even the mandate on which the AHG had been set up. With the most powerful state adopting this position, no agreement was possible. The issue then lay uneasily with states parties in the period from mid-2001 through to the Fifth Review Conference of the BTWC in December 2001. In that period the attacks of September 11 and the subsequent anthrax-impregnated letter attacks generated considerable sympathy for the United States. Some people even wondered if the US might even moderate its position on the BTWC verification protocol.

Despite lingering animosities over the rejection of the AHG's work, the Fifth Review Conference went about its business reasonably successfully – that is, until the very last day. At that point the US representative put forward a proposal that the mandate of the AHG be terminated. This led to such a level of disagreement that no conclusion at all was possible. The Review Conference had to be adjourned for a year. During that year governments such as that of the UK had to consider what could be done to keep the BTWC regime alive. Eventually, at the resumed review in December 2002, it had to be agreed that an Inter-Review-Conference process first proposed by – and therefore acceptable to – the United States should run in the years leading up to the Sixth Review Conference in 2006.

## the inter-review-conference process

At the resumed review the states parties agreed to meet in 2003, 2004 and 2005 to discuss a limited set of topics. Progress in these discussions was then to be reported to the Sixth Review Conference in 2006. The agreement reached for this interim process is shown in Table 9.2 and certain of its features need to be clearly understood. Firstly, agreement could again only be by consensus, so the process was not going to move quickly (particularly in the difficult atmosphere generated by the failure of the protocol negotiations). Secondly, with only a total of three weeks assigned to the topics for each year, it was unlikely that these complex issues could be analysed and decided upon easily.

Still, the aim was clearly set out – "to discuss and promote common understanding and effective action" on the topics set for each year. Furthermore, though it would have been more sensible to agree a protocol first and then deal with these issues, the issues are nevertheless important. Success in reaching agreement, for example on national measures that would be effective in maintaining security and oversight of micro-organisms and toxins, must surely be in everyone's interest, particularly in the longer term as hostile use by sub-state groups becomes more probable. Success in the interim process is also important to set a good foundation for the Sixth Review Conference in 2006 and continue the process of strengthening the convention.

The topics for the first year of the process looked well chosen. States parties should have effective measures to implement the

**Table 9.2** The BTWC Inter-Review Conference process agreed in 2002

1. The conference decides to hold three annual meetings of the states parties of one week duration each year commencing in 2003 until the Sixth Review Conference, to be held not later than the end of 2006, to discuss and promote common understanding and effective action on:

   i. the adoption of necessary national measures to implement the prohibitions set forth in the Convention, including the enactment of penal legislation;
   ii. national mechanisms to establish and maintain the security and oversight of pathogenic micro-organisms and toxins;
   iii. enhancing international capabilities for responding to, investigating and mitigating the effects of cases of alleged use of biological or toxin weapons or suspicious outbreaks of disease;
   iv. strengthening and broadening national and international institutional efforts and existing mechanisms for the surveillance, detection, diagnosis and combating of infectious diseases affecting humans, animals and plants;
   v. the content, promulgation and adoption of codes of conduct for scientists.

2. All meetings, of both experts and states parties, will reach any conclusions or results by consensus.
3. Each meeting of the states parties will be prepared by a two-week meeting of experts. The topics for consideration at each annual meeting of states parties will be as follows: items (i) and (ii) will be considered in 2003; items (iii) and (iv) in 2004; item (v) in 2005. The first meeting will be chaired by a representative of the Eastern Group, the second by a representative of the Group of Non-Aligned and Other States, and the third by a representative of the Western Group.
4. The meetings of experts will prepare factual reports describing their work.
5. The Sixth Review Conference will consider the work of these meetings and decide on any further action.

convention nationally and good security of potential agents *should* be non-contentious. As it turned out, the result of the 2003 meetings was not greeted with great enthusiasm. It appeared that, having been forced to accept the interim process in order to keep multilateral negotiation going in 2002, many states were not minded to move from discussions to common understanding or to effective action. So, although the meeting generated a great deal of information and a short statement saying what would be useful, the states parties were not prepared to agree on what the US had called "deliverables". A draft report wording of "seven 'basic measures' that states parties would agree to undertake on an urgent basis and report to the Sixth Review Conference on progress to date" was not accepted.

Perhaps learning from the 2003 meetings, the 2004 meeting of experts was reported to have been carried out more systematically. There were hopes that this might lead to a more useful outcome of the final meeting of states parties in December 2004, and this does seem to have been the result. There also appeared to be a growing interest in the scientific community in having an effective input to the discussions in 2005 on a code of conduct. So, although in far from good shape, the BTWC regime was not at a dead end in late 2004 and many more NGOs were beginning to ask how the 2006 Sixth Review Conference could be made effective in strengthening the regime.

## conclusions

The story of biological arms control and disarmament told briefly in this chapter is not one of total failure. A convention that supplements the 1925 Geneva Protocol does exist, and over the years there has been a steady growth in the number of states parties acceding to it. On the other hand, the story is far from one of even moderate success. The original BTWC lacked essential components – a verification mechanism and an international organization – and decades of effort have failed to rectify these deficiencies. Furthermore, the world's leading power – which, if it had wished, could probably have carried all the other states parties along to an agreement on improved verification – chose to wreck all the work carried out between the Third Review Conference of 1991 and the Fifth Review Conference of 2001.

The United States still claimed to support the BTWC and the prohibitions it embodies, but it is difficult to think of a more effective way of calling a long-established norm into question. Moreover, other states about which many people have concerns were able to let the US take the blame for the whole failure. One commentator asked openly whether – given this history – the states parties could be considered up to the task of fulfilling their obligations. It certainly seems that much greater general public interest and specialist scientific and medical community attention will be required if the regime is to fare better in the coming decades.

## references

1. Dando, M.R. (1994) *Biological Warfare in the 21st Century: Biotechnology and the Proliferation of Biological Weapons.* London: Brassey's.
2. Dando, M.R. (2002) *Preventing Biological Warfare: The Failure of American Leadership.* Basingstoke, England: Palgrave.
3. See <www.opbw.org>.

chapter ten

# conclusion: the future?

## introduction

In the rest of the world a widespread response to the re-election of President George W. Bush has been despair about the possibility of making progress in dealing co-operatively with multilateral problems over the next four years. Worse still, many commentators wonder if a lasting change has occurred in the United States and the superpower is heading for a protracted period of right-wing government. More urgently, the departure of the moderating influence of Colin Powell as Secretary of State perhaps suggests that the administration's agenda of forcing its view of "democracy" on the world will proceed rapidly in a series of new adventures – maybe against Syria, Iran or North Korea.

Should such a course be followed by the United States over an extended period, it seems likely that the current enormous levels of expenditure in that country on biodefence will continue. Given the lack of transparency, and the obstacles to other intelligence services understanding what is really happening, it is likely that there will be similar expansions of "biodefence" programmes in other countries. With the rate of change in the relevant science and technology and the likelihood of easier terrorist access to dangerous biological materials, states may well begin to see international law as a less and less important element in the protection of their military forces and civilian populations from biowarfare and bioterrorism. At first this may just mean greater reliance on other elements in the web of prevention, but it must be remembered that deterrence through threat of retaliation in kind has a long history and we could end up with the prohibition on the hostile use of modern biology being overturned.

Should an action/reaction arms race break out in biotechnology, which is certain to be one of the major technologies of the twenty-first century, we can discern some of the outlines of what could happen.[1] As we saw in Chapter 7, the problem that the defence eventually faces becomes impossible to deal with. As the century progresses, more and more of life's fundamental processes will become open to manipulation for good – and malevolent – reasons. So the attack could switch from concentration on the agent to concentration on the target. With the huge range of potential targets in the physiological systems of living organisms, and the many ways in which each might be attacked, the advantage will remain for a very long time with the attacker. As argued elsewhere, this would not be a sensible course for international society to follow. The view taken here is that the only way to deal with what will clearly be an escalating threat of the hostile use of biology is through multilateral co-operation aimed at achieving jointly agreed ways to continue strengthening the web of prevention. This seems to be the most likely thing to happen in the coming decades, for the United States will surely return to a more moderate and co-operative position.

The reason for adopting this optimistic view has nothing to do with the long tradition of democracy and good deeds in and by the United States; it arises from the harsh reality of world affairs. Following the collapse of the Soviet Union, the United States found itself the strongest military and economic power in the world. Some in the United States then imagined that this gave them control, rather than great influence, over world affairs. As Robert Jay Lifton has cogently argued, this has created an impossible psychological conundrum: people in leadership aspiring to total control of a situation that, in their hearts, they know they cannot control. In regard to terrorism, Lifton notes,[2] "The world's only superpower has become a target not just because it is so dominant but because its recent policies and attitudes, emerging from superpower syndrome, have antagonized just about everyone." Lifton's view is that the more the United States attempts to exert total control, the more there will be a reaction worldwide against these attempts. This is a view similar to that of Joseph Rotblat (see Chapter 1).

It might be argued that if the United States was sufficiently strong militarily and economically, it could still win any confrontation with the rest of the world. Its military may well be stronger than that of any other country, but its economy is showing clear signs of overextension. The reasons for this are complex, but, as Emmanuel Todd has argued,[3] "Increasingly, the rest of the world is producing so that America can consume. There is no equilibrium between exports and imports establishing itself where the United States is concerned." In 2001, for example, the US trade deficit with China was US $83 billion, with Japan US $68 billion and with the European Union US $60 billion. The US had smaller but nevertheless significant trade deficits with other countries, such as US $30 billion with Mexico and US $13 billion with South Korea. Even worse, with the decline of manufacturing in the United States the nation has little chance to deal with this deficit. The United States, therefore, is basically dependent on the rest of the world. As the economists might say, eventually there will have to be a reckoning.

The view taken here, then, is essentially an optimistic one similar to that eventually reached by Lifton – that the United States is overextended and will have to retrench and, as part of that retrenchment, it will come to embrace more multilateral analyses and solutions to joint problems. Managing the United States' accommodation to reality may not be easy for the rest of us, but it is far from impossible. The task now is not to sink into despair about the failure of the BTWC verification protocol negotiations and the unhelpful position of the United States with regard to strengthening the BTWC, but to find ways to move forward so that we are in a better position to strengthen this central core of the prohibition regime in the future.

## strengthening the web of prevention

The international community has a problem today primarily with state-level offensive biological weapons programmes. These are dangerous because they are the most likely places for major advances resulting from the revolution in the life sciences to be applied in new forms of hostile use. Such programmes are also dangerous because they can lead to other states initiating programmes and then the weapons developed might well be used if all-out conflict were to break out in some region of the world. The international community also has a growing problem with potential bioterrorism. Mass destruction types of attack against people seem beyond the capabilities of terrorists today, but their capabilities will surely grow.

However, lower-level attacks on people and damaging attacks on agriculture are certainly possible today.

Our task is to persuade both states and sub-state groups that the cost of going down the biowarfare or bioterrorism route will far exceed any gain that they will achieve from such efforts. In order to do that we have to make every element of the web of prevention as strong as we possibly can in the coming years and decades – seizing whatever opportunities arise or can be generated at any particular time. A wide range of actions can be undertaken by a very wide range of people. We can all, in various ways, help to make the world a safer place for our children and grandchildren. Every bit of strengthening of the prohibition we achieve helps to reinforce the other elements of the web of prevention and increases the cost to a potential violator of the prohibition. What follows here, therefore, should be viewed as a presentation of some of the current possibilities. There is no doubt that these could be expanded now by the application of other people's experience and imagination, and in the future as we build upon what has already been achieved.

## a web of opportunities

Nicholas Sims, who has written extensively on the history of the BTWC over the last two decades, produced a reflective article after the Fifth Review Conference considering what had gone wrong and what now needed to be done.[4] Though not excusing the United States for the failure of the protocol negotiations, he was at pains to point out that other countries carried a share of the responsibility. China, Cuba, Indonesia, Libya, Pakistan, Sri Lanka and Iran, for example, had issued a common statement in May 2001 which opposed moving on from the rolling text to the composite text that had been produced by the Chairman in an effort to draw the negotiations to a conclusion (see Chapter 9). This common statement was widely seen by supporters of the protocol as very unhelpful, but the countries involved were able to hide behind the United States' outright rejection of the Chairman's text and the negotiation mandate in the July following.

Yet the much larger group of states that did wish to see progress could have moved forward at the original December 2001 Review Conference and the November 2002 resumed session if they had really wished to do so. The usual practice is always to proceed by consensus, but the rules do allow for voting and the intransigence of the United States could have been overcome by a two-thirds majority vote in favour of producing a proper final declaration. This is not as far-fetched as it may seem, for the rest of the world had gone ahead with the Kyoto environmental agreement without the United States. The BTWC Fifth Review Conference Final Declaration was practically complete when the United States employed its last-minute wrecking tactics. If there had been the will in the rest of the international community, a vote could have been taken and work on the protocol might have continued.

We shall return to Sims' analyses later, but first it must be appreciated that many other options are available to the BTWC states parties. The American analyst Jonathan Tucker, for example, noted that in the US presidential election race John Kerry had indicated that he would attempt to renew negotiations to strengthen the bioweapons ban.[5] Tucker, who believes that the envisaged protocol was not a sensible proposal, nevertheless suggested one option – that America could try to move forward by taking a "building blocks" approach, by initiating negotiation of elements of the protocol which might be simpler to agree separately. These are the elements:

> The *Biosecurity Protocol* would establish a set of functional standards for the physical protection, control, and accounting of dangerous pathogens and toxins ...
>
> The *Investigation Protocol* would contain provisions for conducting field investigations of the alleged use of BW agents and suspicious outbreaks of infectious disease ...
>
> The *Inspection Protocol* would set out procedures for "challenge-type" visits, at the request of a state party, to any development, production, or storage facility on the territory of another member state that is suspected of a treaty violation.

Following President Bush's re-election, such an approach seems unlikely at present, but that does not mean it may not be considered again in a few years' time.

Even if we accept that any action related to the protocol is impossible at present, many things could still be done now by states parties to reinforce elements of the web of prevention. This is well illustrated by the efforts of the United Kingdom Foreign Office in the period between the adjournment of the Fifth Review Conference in 2001 and the renewed session in late 2002. In April 2002 the UK produced a Green Paper entitled *Strengthening the Biological and Toxin Weapons Convention: Countering the Threat from Biological Weapons*.[6] This paper included a list of possible options that might be taken forward. For example, in its background paper on scientific and technological developments relevant to the convention for the 2001 review, the United Kingdom had questioned whether it was prudent to review such developments only at five-yearly intervals in view of the accelerating pace of scientific and technological change in the life sciences. In the Green Paper this viewpoint was considerably strengthened by the suggestion that the accelerating pace of change made it "quite unsafe" to have only the five-yearly reviews. The Green Paper suggested setting up a Scientific Advisory Panel which would meet every one or two years to consider and assess developments and their implications for the convention and for measures being taken to strengthen it.

Graham Pearson, who has reported in detail on the BTWC negotiations in Geneva, reviewed both the United Kingdom's proposals and proposals put forward by other states parties at the Fifth Review Conference. He was then able to assemble a list of proposals and to arrange them in what he considered an ascending order of difficulty for negotiation by the states parties.[7] Table 10.1 reproduces the headings from Pearson's listing in the order of increasing amount of negotiation required. It is obvious from this listing that there is much that can and should be done to strengthen the web of prevention.

A case in point here is that, although scientific and technological changes are taking place just as fast in animal and plant biology as in human biology, in discussions on biological warfare and bioterrorism the understandable focus on human pathogens prevents proper consideration of attacks on agriculture. A working paper from South Africa at the Fifth Review Conference considered how changes to the agreed Confidence-Building Measures might help to deal with this problem. It suggested specific modifications to CBM A on maximum containment facilities and CBM G on vaccine production facilities so that they included animal and plant pathogen facilities in addition to human pathogen facilities. It also proposed a new CBM H requiring declaration of plant inoculant and biocontrol agent production facilities. If agreed and enacted, such CBMs would greatly assist in giving transparency to national activities where there is very rapid change (and where there are obvious possibilities for misuse).

**Table 10.1** Measures proposed to strengthen the convention*

- Actively promoting universal membership of the BTWC
- Withdrawal of reservations to the 1925 Geneva Protocol
- Increase efforts on disease surveillance, detection and diagnosis and countering infectious disease generally
- Code of conduct for professional bodies
- National criminal legislation and extradition
- Enhanced national controls on dangerous pathogens
- Oversight of genetic engineering and high-risk experiments
- Revised CBMs with specific extensions to particular CBMs
- Regular annual meetings of subsidiary bodies such as a Scientific Advisory Panel and/or an Oversight Committee
- Assistance in the event of, or threat of, use of BW
- A new Convention on Criminalization of CBW
- A new Convention on Physical Protection of Dangerous Pathogens
- Article X implementation
- Investigations into non-compliance (alleged use, misuse of facilities, suspicious outbreaks) and effective compliance machinery to make it much harder to cheat
- Guidelines to ensure strengthening of Article III and to prohibit transfers of dual-use materials to non-state actors
- Voluntary annual notification of authorized transfers
- Use of Cartagena Biosafety Protocol advanced informed agreement provisions for transfer of living modified organisms

* From Pearson, op. cit.

As we know, the resumed Fifth Review Conference agreed in late 2002 to consider a restricted set of topics over the period 2003–5 and decided that the results of deliberations on these topics would only be considered at the 2006 Sixth Review Conference. What is more, this process and these topics were chosen to ensure that the United States was unlikely to walk away from the deliberations. What the net outcome of this new process will be in 2006 remains to be seen. It is clear, however, that the process is an *interim measure* agreed in difficult circumstances and cannot and should not be seen as a substitute for the regular Five-Year Review process of developing the BTWC.

As Sims pointed out at the end of his reflections on the Fifth Review Conference referred to earlier, there are good reasons why multilateral diplomacy works. These can be summarized as follows:[8]

> It promotes universality through the search for solutions which are tolerable (if not optimal) for as many participants as possible.
>
> It steers the evolution of treaty regimes in an acceptable way, because everyone can feel involved, and common understandings about norms and expectations have a basis not just in consent but in shared formulation.
>
> It reaffirms such basic structural elements as mutuality, reciprocity and equality of obligation within existing and prospective treaty regimes, thereby emphasising co-operation and consultation, with compliance concerns more likely to be addressed in problem-solving than in adversarial mode.

The main mechanism to carry forward the task of strengthening the BTWC is the regular Five-Year Review. To meet the requirements of the second point in the list above there has therefore to be effective review and agreement at these meetings. And there is much that needs to be done besides agreement of a protocol, for, as Sims points out, there is still no organization to support the BTWC like the Organization for the Prohibition of Chemical Weapons that takes care of the CWC. In Sims's words, "The B[T]WC has no annual assembly, no governing council, no standing committees or advisory panels, not even a permanent secretariat" – and this is a convention agreed some thirty years ago to control what is increasingly viewed as a critical threat to international security!

## grasping the opportunities

In order to deal effectively with the issues raised in this book, there must be an effective input from the relevant scientific community. The community of life scientists has to own the problem of the potential misuse of their work in the same way as do the many physicists who have recognized their responsibilities since Hiroshima and Nagasaki. Life scientists' concern and effective input have been largely lacking in the past decades. Few biomedical scientists knew that an attempt was being made to strengthen the BTWC during the 1990s despite the considerable efforts of such organizations as Pugwash, the International Committee of the Red Cross and the British Medical Association.[9]

That gap in knowledge is likely to be closed because the topic of discussion in 2005 in the new BTWC process is to be "Codes of Conduct for Scientists". Already in the United States we have seen a National Academies report on *Biosecurity in an Age of Terrorism*, suggesting new controls on the operations of the life sciences community, and this suggestion has been taken up by the governing administration. It seems unlikely that politicians will be willing to let the revolution in biology proceed without further consideration of potential misuse. So the scientists will have to become more active and involved and bring their expertise to bear on intelligent ways to develop the web of prevention.

International civil society, for its part, is beginning to become more organized in regard to preventing the hostile use of biology. Though the kind of large-scale, informed, non-governmental presence that focuses on nuclear weapons or landmines is some way distant, the joint Bioweapons Prevention Project (BWPP) is now entering its third year of operation and is steadily expanding the worldwide network of active organizations.

What is still missing is an equivalent network of states parties determined to take the lead in getting effective action. That is where readers of this book come in. Whatever their view of the strengths and weaknesses of the arguments presented in the foregoing chapters, it is not in their or their family's interest to let the current low level of awareness continue. We need a widespread, informed public debate on what needs to be done to prevent biowarfare and bioterrorism in the years ahead. If there is no such informed debate there will be no pressure for effective action at the national political level – unless, of course, there is some catastrophe, in which case knee-jerk reactions and bad policy are the likely result. In our different countries we have to raise the issue in a coherent, consistent way. We have to break out of the fragmented, disjointed perceptions that arise from the current intermittent media coverage and develop the capability to ask our political representatives the critical questions that will motivate them to seek answers and changed behaviour from our governments.

Without this effort from many, many people, the present policy of drift is likely to continue. That policy response to the danger of biological weapons development and use in the years to come led one careful observer of the Geneva negotiations to ask whether "the States Parties as a collective body are actually up to the tasks they are legally bound to undertake".[10] If we do allow the new life sciences to be applied in a major way to hostile purposes – as all previous revolutions in science and technology have been applied – the future will be grim. Once Pandora's box is open, it will be almost impossible to close.

## references


1. Petro, J.B., Plasse, T.R. and McNulty, J.A. (2003) Biotechnology: impact on biological warfare and biodefense. *Biosecurity and Bioterrorism*, **1**(3), 161–8.
2. Lifton, R.J. (2003) *Superpower Syndrome: America's Apocalyptic Confrontation with the World.* New York: Nation Books.
3. Todd, E. (2004) *After the Empire: The Breakdown of the American Order.* London: Constable.
4. Sims, N.A. (2003) Biological disarmament diplomacy in the doldrums: reflections after the BWC Fifth Review Conference. *Disarmament Diplomacy*, April/May, 11–18.
5. Tucker, J.B. (2004) Strengthening the BWC: a way forward. *Disarmament Diplomacy*, July/August, 24–30.
6. United Kingdom (2002) *Strengthening the Biological and Toxin Weapons Convention: Countering the Threat from Biological Weapons.* London: Stationery Office.
7. Pearson, G.S. (2002) *Return to Geneva: A Comprehensive List of Measures.* Bradford: Department of Peace Studies, University of Bradford.
8. Sims, op. cit.
9. Dando, M.R. and Nathanson, V. (2004) *Biotechnology, Weapons and Humanity II.* London: British Medical Association.
10. Littlewood, J. (2003) *Preparing for a Successful Outcome to the BTWC Sixth Review Conference in 2006.* Bradford: Department of Peace Studies, University of Bradford.

appendix 1

# appeal on biotechnology, weapons and humanity

## background

The "age of biotechnology", like the industrial revolution and the "information age", promises great benefits to humanity. Yet if biotechnology is put to hostile uses, including to spread terror, the human species faces great dangers.

The International Committee of the Red Cross (ICRC), in keeping with its mandate to protect and assist victims of armed conflict, is particularly alarmed by the potential hostile uses of biological agents.

Potential benefits of advances in biological sciences and technologies are impressive. These include cures for diseases, new vaccines and increases in food production, including in impoverished regions of the world.

Yet the warnings of what can go wrong are profoundly disturbing. The ICRC believes these merit reflection at every level of society. Testimony from governments, UN agencies, scientific circles, medical associations and industry provides a long list of existing and emerging capacities for misuse. These include:

- Deliberate spread of existing diseases such as typhoid, anthrax and smallpox to cause death, disease and fear in a population.
- Alteration of existing disease agents rendering them more virulent, as already occurred unintentionally in research on the "mousepox" virus.

- Creation of viruses from synthetic materials, as occurred this year using a recipe from the Internet and gene sequences from a mail order supplier.
- Possible future development of ethnically or racially specific biological agents.
- Creation of novel biological warfare agents for use in conjunction with corresponding vaccines for one's own troops or population. This could increase the attractiveness of biological weapons.
- New methods to covertly spread naturally occurring biological agents to alter physiological or psychological processes of target populations such as consciousness, behavior and fertility, in some cases over a period of years.
- Production of biological agents that could attack agricultural or industrial infrastructure. Even unintended release of such agents could have uncontrollable and unknown effects on the natural environment.
- Creation of biological agents that could affect the makeup of human genes, pursuing people through generations and adversely affecting human evolution itself.

The life processes at the core of human existence must never be manipulated for hostile ends. In the past, scientific advances have all too often been misused. It is essential that humanity acts together now to prevent the abuse of biotechnology.

The ICRC calls on all concerned to assume their responsibilities in this field, before it is too late. We must reaffirm the ancient taboo against the use in war of "plague and poison", passed down for generations in diverse cultures. From the ancient Greeks and Romans, to the Manu Law of War in India, to rules on the conduct of war drawn from the Koran by the Saracens, the use of poison and poison weapons has been forbidden. This ban was codified in the 1863 Lieber Code during the US Civil War and, internationally, in the 1899 Hague Declaration and the Regulations annexed to the 1907 Hague Convention IV.

In February 1918, the ICRC launched an impassioned appeal, describing warfare by poison as "a barbaric invention which science is bringing to perfection ..." and protesting "with all the force at [its] command against such warfare, which can only be called criminal." This appeal is still valid today.

Responding in part to the ICRC's appeal, States adopted the 1925 Geneva Protocol, reaffirming the general ban on the use of poison gas and extending it to cover bacteriological weapons. This norm is now part of customary international law – binding on all parties to all armed conflicts.

The 1972 Biological Weapons Convention significantly reinforced this prohibition by outlawing the development, production, stockpiling, acquisition, retention and transfer of biological weapons. As regards new advances in biotechnology and possible terrorist threats, this Convention covers *all biological agents* which "have no justification for prophylactic, protective or other peaceful purposes" and includes the means to deliver such agents. (Article 1, 1972 Biological Weapons Convention). The ICRC deeply regrets that lengthy negotiations to strengthen this Convention through a compliance-monitoring regime did not come to fruition as expected in November 2001. This underlines the urgent need for a renewed commitment by all States to ensure effective control of biological agents.

The responsibility to prevent hostile uses of biotechnology lies with each State. But it extends beyond governments to all persons, especially to military, scientific and medical professionals and those in the biotechnology and pharmaceutical industries.

## full text:

## appeal of the international committee of the red cross on biotechnology, weapons and humanity

Alarmed by the potential hostile uses of biotechnology, *the International Committee of the Red Cross (ICRC) appeals* to:

- all political and military authorities to strengthen their commitment to the international humanitarian law norms which prohibit the hostile uses of biological agents, and to work together to subject potentially dangerous biotechnology to effective controls.
- the scientific and medical communities, industry and civil society in general to ensure that potentially dangerous biological knowledge and agents be subject to effective controls.

## the ICRC appeals in particular:

### to all political and military authorities

- To become parties to the 1925 Geneva Protocol and the 1972 Biological Weapons Convention, if they have not already done so, to encourage States which are not parties to become parties, and to lift reservations on use to the 1925 Geneva Protocol,
- To resume with determination efforts to ensure faithful implementation of these treaties and develop appropriate mechanisms to maintain their relevance in the face of scientific developments,
- To adopt stringent national legislation, where it does not yet exist, for implementation of the 1925 Geneva Protocol and the 1972 Biological Weapons Convention, and to enact effective controls on biological agents with potential for abuse,
- To ensure that any person who commits acts prohibited by the above instruments is prosecuted,
- To undertake actions to ensure that the legal norms prohibiting biological warfare are known and respected by members of armed forces,
- To encourage the development of effective codes of conduct by scientific and medical associations and by industry to govern activities and biological agents with potential for abuse, and
- To enhance international cooperation, including through the development of greater international capacity to monitor and respond to outbreaks of infectious disease.

### to the scientific and medical communities and to the biotechnology and pharmaceutical industries

- To scrutinize all research with potentially dangerous consequences and to ensure it is submitted to rigorous and independent peer review,
- To adopt professional and industrial codes of conduct aimed at preventing the abuse of biological agents,
- To ensure effective regulation of research programs, facilities and biological agents which may lend themselves to misuse, and supervision of individuals with access to sensitive technologies, and
- To support enhanced national and international programs to prevent and respond to the spread of infectious disease.

The ICRC calls on all those addressed here to assume their responsibilities as members of a species whose future may be gravely threatened by abuse of biological knowledge. The ICRC appeals to you to make your contribution to the age-old effort to protect humanity from disease. We urge you to consider the threshold at which we all stand and to remember our common humanity.

The ICRC urges States to adopt at a high political level an international Declaration on "Biotechnology, Weapons and Humanity" containing a renewed commitment to existing norms and specific commitments to future preventive action.

Geneva, September 2002

# the biological and toxin weapons convention

Convention on the Prohibition of the Development, Production and Stockpiling of Bacteriological (Biological) and Toxin Weapons and on Their Destruction.

Signed at London, Moscow and Washington on 10 April 1972.
Entered into force on 26 March 1975.
Depositaries: U.K., U.S. and Soviet governments.

The States Parties to this Convention,

Determined to act with a view to achieving effective progress towards general and complete disarmament, including the prohibition and elimination of all types of weapons of mass destruction, and convinced that the prohibition of the development, production and stockpiling of chemical and bacteriological (biological) weapons and their elimination, through effective measures, will facilitate the achievement of general and complete disarmament under strict and effective international control,

Recognizing the important significance of the Protocol for the Prohibition of the Use in War of Asphyxiating, Poisonous or Other Gases, and of Bacteriological Methods of Warfare, signed at Geneva on June 17, 1925, and conscious also of the contribution which the said Protocol has already made, and continues to make, to mitigating the horrors of war,

Reaffirming their adherence to the principles and objectives of that Protocol and calling upon all States to comply strictly with them,

Recalling that the General Assembly of the United Nations has repeatedly condemned all actions contrary to the principles and objectives of the Geneva Protocol of June 17, 1925,

Desiring to contribute to the strengthening of confidence between peoples and the general improvement of the international atmosphere,

Desiring also to contribute to the realization of the purposes and principles of the United Nations,

Convinced of the importance and urgency of eliminating from the arsenals of States, through effective measures, such dangerous weapons of mass destruction as those using chemical or bacteriological (biological) agents,

Recognizing that an agreement on the prohibition of bacteriological (biological) and toxin weapons represents a first possible step towards the achievement of agreement on effective measures also for the prohibition of the development, production and stockpiling of chemical weapons, and determined to continue negotiations to that end,

Determined for the sake of all mankind, to exclude completely the possibility of bacteriological (biological) agents and toxins being used as weapons,

Convinced that such use would be repugnant to the conscience of mankind and that no effort should be spared to minimize this risk,

Have agreed as follows:

## article I

Each State Party to this Convention undertakes never in any circumstances to develop, produce, stockpile or otherwise acquire or retain:

(1) Microbial or other biological agents, or toxins whatever their origin or method of production, of types and in quantities that have no justification for prophylactic, protective or other peaceful purposes;

(2) Weapons, equipment or means of delivery designed to use such agents or toxins for hostile purposes or in armed conflict.

## article II

Each State Party to this Convention undertakes to destroy, or to divert to peaceful purposes, as soon as possible but not later than nine months after entry into force of the Convention, all agents, toxins, weapons, equipment and means of delivery specified in article I of the Convention, which are in its possession or under its jurisdiction or control. In implementing the provisions of this article all necessary safety precautions shall be observed to protect populations and the environment.

## article III

Each State Party to this Convention undertakes not to transfer to any recipient whatsoever, directly or indirectly, and not in any way to assist, encourage, or induce any State, group of States or international organizations to manufacture or otherwise acquire any of the agents, toxins, weapons, equipment or means of delivery specified in article I of this Convention.

## article IV

Each State Party to this Convention shall, in accordance with its constitutional processes, take any necessary measures to prohibit and prevent the development, production, stockpiling, acquisition, or retention of the agents, toxins, weapons, equipment and means of delivery specified in article I of the Convention, within the territory of such State, under its jurisdiction or under its control anywhere.

## article V

The States Parties to this Convention undertake to consult one another and to cooperate in solving any problems which may arise in relation to the objective of, or in the application of the provisions of, the Convention. Consultation and Cooperation pursuant to this article may also be undertaken through appropriate international procedures within the framework of the United Nations and in accordance with its Charter.

## article VI

(1) Any State Party to this convention which finds that any other State Party is acting in breach of obligations deriving from the provisions of the Convention may lodge a complaint with the Security Council of the United Nations. Such a complaint should include all possible evidence confirming its validity, as well as a request for its consideration by the Security Council.

(2) Each State Party to this Convention undertakes to cooperate in carrying out any investigation which the Security Council may initiate, in accordance with the provisions of the Charter of the United Nations, on the basis of the complaint received by the Council. The Security Council shall inform the States Parties to the Convention of the results of the investigation.

## article VII

Each State Party to this Convention undertakes to provide or support assistance, in accordance with the United Nations Charter, to any Party to the Convention which so requests, if the Security Council decides that such Party has been exposed to danger as a result of violation of the Convention.

## article VIII

Nothing in this Convention shall be interpreted as in any way limiting or detracting from the obligations assumed by any State under the Protocol for the Prohibition of the Use in War of Asphyxiating, Poisonous or Other Gases, and of Bacteriological Methods of Warfare, signed at Geneva on June 17, 1925.

## article IX

Each State Party to this Convention affirms the recognized objective of effective prohibition of chemical weapons and, to this end, undertakes to continue negotiations in good faith with a view to reaching early agreement on effective measures for the prohibition of their development, production and stockpiling and for their destruction, and on appropriate measures concerning equipment and means of delivery specifically designed for the production or use of chemical agents for weapons purposes.

## article X

(1) The States Parties to this Convention undertake to facilitate, and have the right to participate in, the fullest possible exchange of equipment, materials and scientific and technological information for the use of bacteriological (biological) agents and toxins for peaceful purposes. Parties to the Convention in a position to do so shall also cooperate in contributing individually or together with other States or international organizations to the further development and application of scientific discoveries in the field of bacteriology (biology) for prevention of disease, or for other peaceful purposes.

(2) This Convention shall be implemented in a manner designed to avoid hampering the economic or technological development of States Parties to the Convention or international cooperation in the field of peaceful bacteriological (biological) activities, including the international exchange of bacteriological (biological) agents and toxins and equipment for the processing, use or production of bacteriological (biological) agents and toxins for peaceful purposes in accordance with the provisions of the Convention.

## article XI

Any State Party may propose amendments to this Convention. Amendments shall enter into force for each State Party accepting the amendments upon their acceptance by a majority of the States Parties to the Convention and thereafter for each remaining State Party on the date of acceptance by it.

## article XII

Five years after the entry into force of this Convention, or earlier if it is requested by a majority of Parties to the Convention by submitting a proposal to this effect to the Depositary Governments, a conference of States Parties to the Convention shall be held at Geneva, Switzerland, to review the operation of the Convention, with a view to assuring that the purposes of the preamble and the provisions of the Convention, including the provisions concerning negotiations on chemical weapons, are being realized. Such review shall take into account any new scientific and technological developments relevant to the Convention.

## article XIII

(1) This Convention shall be of unlimited duration.

(2) Each State Party to this Convention shall in exercising its national sovereignty have the right to withdraw from the Convention if it decides that extraordinary events, related to the subject matter of the Convention, have jeopardized the supreme interests of its country. It shall give notice of such withdrawal to all other States Parties to the Convention and to the United Nations Security Council three months in advance. Such notice shall include a statement of the extraordinary events it regards as having jeopardized its supreme interests.

## article XIV

(1) This Convention shall be open to all States for signature. Any State which does not sign the Convention before its entry into force in accordance with paragraph (3) of this Article may accede to it at any time.

(2) This Convention shall be subject to ratification by signatory States. Instruments of ratification and instruments of accession shall be deposited with the Governments of the United States of America, the United Kingdom of Great Britain and Northern Ireland and the Union of Soviet Socialist Republics, which are hereby designated the Depositary Governments.

(3) This Convention shall enter into force after the deposit of instruments of ratification by twenty-two Governments, including the Governments designated as Depositaries of the Convention.

(4) For States whose instruments of ratification or accession are deposited subsequent to the entry into force of this Convention, it shall enter into force on the date of the deposit of their instruments of ratification or accession.

(5) The Depositary Governments shall promptly inform all signatory and acceding States of the date of each signature, the date of deposit of each instrument of ratification or of accession and the date of entry into force of this Convention, and of the receipt of other notices.

(6) This Convention shall be registered by the Depositary Governments pursuant to Article 102 of the Charter of the United Nations.

## article XV

This Convention, the English, Russian, French, Spanish and Chinese texts of which are equally authentic, shall be deposited in the archives of the Depositary Governments. Duly certified copies of the Convention shall be transmitted by the Depositary Governments to the Governments of the signatory and acceding states.

# index

Page numbers in *italics* represent figures and tables.